W9-BHV-594

CHEMICAL SENSORS AND BIOSENSORS

Analytical Techniques in the Sciences (AnTS)

Series Editor: David J. Ando, Consultant, Dartford, Kent, UK

A series of open learning/distance learning books which covers all of the major analytical techniques and their application in the most important areas of physical, life and materials science.

Titles Available in the Series

Analytical Instrumentation: Performance Characteristics and Quality
Graham Currell, University of the West of England, Bristol, UK

Fundamentals of Electroanalytical Chemistry
Paul M. S. Monk, Manchester Metropolitan University, Manchester, UK

Introduction to Environmental Analysis
Roger N. Reeve, University of Sunderland, UK

Polymer Analysis
Barbara H. Stuart, University of Technology, Sydney, Australia

Chemical Sensors and Biosensors
Brian R. Eggins, University of Ulster at Jordanstown, Northern Ireland, UK

Forthcoming Titles

Analysis of Controlled Substances
Michael D. Cole, Anglia Polytechnic University, Cambridge, UK

Liquid Chromatography–Mass Spectrometry: An Introduction
Robert E. Ardrey, University of Huddersfield, UK

CHEMICAL SENSORS AND BIOSENSORS

Brian R. Eggins
University of Ulster at Jordanstown
Northern Ireland, UK

JOHN WILEY & SONS, LTD

Baffins Lane, Chichester,
West Sussex, PO19 1UD, England

National 01243 779777
International (+44) 1243 779777

e-mail (for orders and customer service enquiries): cs-books@wiley.co.uk
Visit our Home Page on http://www.wileyeurope.com or http://www.wiley.com

Other Wiley Editorial Offices

John Wiley & Sons, Inc., 605 Third Avenue,
New York, NY 10158-0012, USA

Wiley-VCH Verlag GmbH,
Pappelallee 3, D-69469 Weinheim, Germany

John Wiley & Sons Australia, Ltd
33 Park Road, Milton, Queensland 4064, Australia

John Wiley & Sons (Asia) Pte Ltd, 2 Clementi Loop #02-01,
Jin Xing Distripark, Singapore 129809

John Wiley & Sons (Canada) Ltd, 22 Worcester Road,
Rexdale, Ontario M9W 1L1, Canada

Library of Congress Cataloging-in-Publication Data

Eggins, Brian R.
Chemical sensors and biosensors / Brian R. Eggins.
p. cm. – (Analytical techniques in the sciences (AnTS))
Includes bibliographical references and index.
ISBN 0-471-89913-5 (cloth: alk. paper) – ISBN 0-471-89914-3 (pbk. : alk. paper)
1. Chemical detectors. 2. Biosensors. I. Title. II. Analytical techniques in the sciences.

TP159.C46 E44 2002
660′.283 – dc21 2001057388

British Library Cataloguing in Publication Data

A catalogue record for this book is available from the British Library

ISBN 0-471-89913-5 (Cloth)
ISBN 0-471-89914-3 (Paper)

Typeset in 10/12pt Times by Laserwords Private Limited, Chennai, India.
Printed and bound in Great Britain by Antony Rowe, Chippenham, Wiltshire.
This book is printed on acid-free paper responsibly manufactured from sustainable forestry in which at least two trees are planted for each one used for paper production.

To Chrissie and Rosanne

Contents

Series Preface

There has been a rapid expansion in the provision of further education in recent years, which has brought with it the need to provide more flexible methods of teaching in order to satisfy the requirements of an increasingly more diverse type of student. In this respect, the *open learning* approach has proved to be a valuable and effective teaching method, in particular for those students who for a variety of reasons cannot pursue full-time traditional courses. As a result, John Wiley & Sons Ltd first published the Analytical Chemistry by Open Learning (ACOL) series of textbooks in the late 1980s. This series, which covers all of the major analytical techniques, rapidly established itself as a valuable teaching resource, providing a convenient and flexible means of studying for those people who, on account of their individual circumstances, were not able to take advantage of more conventional methods of education in this particular subject area.

Following upon the success of the ACOL series, which by its very name is predominately concerned with Analytical *Chemistry*, the *Analytical Techniques in the Sciences* (AnTS) series of open learning texts has now been introduced with the aim of providing a broader coverage of the many areas of science in which analytical techniques and methods are now increasingly applied. With this in mind, the AnTS series seeks to provide a range of books which will cover not only the actual techniques themselves, but *also* those scientific disciplines which have a necessary requirement for analytical characterization methods.

Analytical instrumentation continues to increase in sophistication, and as a consequence, the range of materials that can now be almost routinely analysed has increased accordingly. Books in this series which are concerned with the *techniques* themselves will reflect such advances in analytical instrumentation, while at the same time providing full and detailed discussions of the fundamental concepts and theories of the particular analytical method being considered. Such books will cover a variety of techniques, including general instrumental analysis, spectroscopy, chromatography, electrophoresis, tandem techniques, electroanalytical methods, X-ray analysis and other significant topics. In addition, books in the series will include the *application* of analytical techniques in areas such as environmental science, the life sciences, clinical analysis, food science, forensic analysis, pharmaceutical science, conservation and archaeology, polymer science and general solid-state materials science.

Written by experts in their own particular fields, the books are presented in an easy-to-read, user-friendly style, with each chapter including both learning objectives and summaries of the subject matter being covered. The progress of the reader can be assessed by the use of frequent self-assessment questions (SAQs) and discussion questions (DQs), along with their corresponding reinforcing or remedial responses, which appear regularly throughout the texts. The books are thus eminently suitable both for self-study applications and for forming the basis of industrial company in-house training schemes. Each text also contains a large amount of supplementary material, including bibliographies, lists of acronyms and abbreviations, and tables of SI Units and important physical constants, plus where appropriate, glossaries and references to original literature sources.

It is therefore hoped that this present series of textbooks will prove to be a useful and valuable source of teaching material, both for individual students and for teachers of various science courses.

Dave Ando
Dartford, UK

Preface

This book is derived partly from the author's earlier book, *Biosensors: An Introduction*, originally published in 1996. Much of the same material is used here, although now it is set in the style of an open learning book, following the presentation of the Analytical Chemistry by Open Learning (ACOL) Series and now set in the style of the new Analytical Techniques in the Sciences (AnTS) Series. The scope of the previous book is broadened to cover chemical sensors as well as biosensors.

The original *Biosensors* book evolved out of a lecture course in biosensors (given at the University of Ulster), as there were very few suitable textbooks available at that time. A number of part-time students were unable to attend the formal lectures, but used the *Biosensors* textbook in open learning mode, helped by informal tutorials. These students, on average, performed at least as well in the examinations as the corresponding full-time students.

After an introductory chapter which describes the general idea of sensors, Chapter 2 discusses in some detail different transduction elements, both electrochemical and photometric. Chapter 3 then describes the various sensing elements used to select particular analytes, while Chapter 4 discusses performance factors such as selectivity, sensitivity, range, lifetimes, etc. Chapter 5 next describes in more detail the applications of electrochemistry in sensors and biosensors. Photometric sensors are described in Chapter 6, with mass-sensitive and thermal-sensitive sensors being discussed in Chapter 7. Finally, Chapter 8 gives five case studies of particular applications in more detail.

I would like to thank my colleagues at the University of Ulster in the Biomedical Environmental Sensor Technology (BEST) Centre for their help and encouragement, Professor John Anderson, Professor Jim McLaughlin, Dr Tony Byrne and Dr Eric McAdams (at UU Jordanstown), Professors Dermot Diamond, Johannes Vos and Malcolm Smyth (at Dublin City University), and from The J Fourier University in Grenoble, France, Drs Pascal Mailley and Serge Cosnier.

I would like to acknowledge the indirect inspiration I have received through the work and contacts with Professor George Guillbault at University College, Cork, Professor Anthony Turner at Cranfield University, and not least, Professor Allen Hill of Oxford University, whose lecture inspired my first interest in writing books on biosensors. I must also mention, through the 'Eirelec Conferences', Professor Joe Wang from the University of New Mexico, and Professor Allen Bard from the University of Texas at Austin. I would also like to acknowledge the contributions of some of my former students, namely Dr Edward Cummings, Dr Min Zhou, Mr Shane McFadden, Ms Catriona Hickey and Mr Stephen Toft.

I wish particularly to thank David Ando, the Managing Editor of the AnTS Series, for his great help and encouragement in advising and editing the manuscript.

Above all, I thank my wife Chrissie, without whose dedicated support, encouragement and domestic provisions I would never have completed this book.

Brian R. Eggins
University of Ulster at Jordanstown

Acronyms, Abbreviations and Symbols

A	analyte
A*	analyte analogue
Ab	antibody
AC	alternating current
Ag	antigen
AMP	adenosine 5′-monophosphate
AP	acid phosphatase; action potential
ATP	adenosine triphosphate
ATR	attenuated total reflectance
BOD	biological oxygen demand
C	coulomb
CHEMFET	field-effect transistor, with chemically sensing gate
Cp	cyclopentadiene
CPE	carbon paste electrode
CV	cyclic voltammetry
DAC	*p*-dimethylaminocinnamaldeyde
DC	direct current
DEAE	diethylaminoethyl
DNA	deoxyribonucleic acid
DVM	digital voltmeter
ECQM	electrochemical quartz crystal microbalance
emf	electromotive force
ENFET	field-effect transistor, with enzyme gate system
EW	evanescent wave
FAD	flavin–adenine dinucleotide
Fc	ferrocene

FET	field-effect transistor
FIA	flow-injection analysis
FITC	fluorescein isothiocyanate
FMH	flavin mononucleotide
GC	gas chromatography
GDH	glucose dehydrogenase
GOD	glucose oxidase
HMDE	hanging-mercury-drop electrode
HPLC	high performance liquid chromatography
id	internal diameter
IGFET	insulated-gate field-effect transistor
IR	infrared
IRE	internal reflection element
ISA	ionic-strength adjuster
ISE	ion-selective electrode
ISFET	ion-selective field-effect transistor
IUPAC	International Union of Pure and Applied Chemistry
J	joule
LDH	lactate dehydrogenase
LDV	laser Doppler velocimetry
LED	light-emitting diode
LMO	lactate monooxidase
LOD	lactate oxidase
LSV	linear-sweep voltammetry
M	molarity (mol dm^{-3})
MIS	metal–insulator–semiconductor
mM	millimolar (10^{-3} mol dm^{-3})
MS	mass spectrometry
NAD	nicotinamide–adenine dinucleotide
NAD^+	nicotinamide–adenine dinucleotide (oxidized form)
NADH	nicotinamide–adenine dinucleotide (reduced form)
NMP	*N*-methylphenothiazine
Ox	oxidized species
PCS	phase correlation spectroscopy
PO	peroxidase
ppm	parts per million
PPY	polypyrrole
QCM	quartz crystal microbalance
QELS	quasi-elastic light scattering spectroscopy
R	reduced species
RF	radiofrequency
RNA	ribonucleic acid
SAW	surface acoustic wave

SCE	saturated-calomel electrode
SHE	standard hydrogen electrode
SI (units)	Système International (d'Unitès) (International System of Units)
SPE	screen-printed electrode
SPR	surface plasmon resonance
TCNQ	tetracyanoquinodimethane
TIRF	total internal reflection fluorescence
TISAB	total-ionic-strength adjustment buffer
TNT	trinitrotoluene
TTF	tetrathiofulvalene
UV	ultraviolet
V	volt
vis	visible
A	absorbance
C	concentration; capacitance
f	frequency
I	electric current; intensity (of light)
L	conductance
m	mass
R	resistance; molar gas constant
t	time
T	thermodynamic temperature
V	electric potential
ε	extinction coefficient
λ	wavelength
ν	frequency (of radiation)

About the Author

Brian Eggins

The author was educated at King Edward's (Five Ways) School, Birmingham, and at Gonville and Caius College, Cambridge University (B.A. and M.A.). He then obtained an M.Sc. degree at The University of Manchester Institute of Science and Technology (UMIST), followed by a Ph.D. at Warwick University. He was then a Research Associate at Colorado University, USA for two years. After a brief period in industry and teaching experience at Grimsby College of Technology, he moved to Ulster Polytechnic (now The University of Ulster).

He is now a Reader in Physical and Analytical Chemistry at the University of Ulster. His research interests include electrochemistry and photo-electrochemistry, as well as biosensors. He has supervised 15 research students and is currently involved in three EU-funded research projects involving groups from four European countries, as well as from Israel. He has lectured in Canada, the USA and Europe. He has published over 100 original research papers, plus is author of *Chemical Structure and Reactivity,* published by Macmillan, *Estructura Quimica y Reactividad*, published by Ediciones Bellaterra, SA, and *Biosensors: An Introduction,* published jointly by John Wiley and Sons, Ltd and B.G. Teubner. He is a Fellow of The Royal Society of Chemistry.

Chapter 1
Introduction

Learning Objectives

- To define different types of sensors.
- To list recognition elements.
- To list the transducers used in sensors.
- To learn the methods of attaching recognition elements to transducers.
- To understand the most important performance factors.
- To know three main areas of application.

1.1 Introduction to Sensors

1.1.1 What are Sensors?

DQ 1.1

What is a sensor?

Answer

We have at least five of these, i.e. our noses, our tongues, our ears, our eyes and our fingers. They represent the main types of sensor. In the laboratory, one of the best known types of sensor is the litmus paper test for acids and alkalis, which gives a qualitative indication, by means of a colour reaction, of the presence or absence of an acid. A more precise method of indicating the degree of acidity is the measurement of pH, either by the more extended use of colour reactions in special indicator solutions, or even by simple pH papers. However, the best method of measuring acidity is the use of the pH meter, which is an electrochemical device giving an electrical response which can be read by a needle moving on a scale or on a digital read-out device or input to a microprocessor.

In such methods, the *sensor* that responds to the degree of acidity is either a chemical – the dye litmus, or a more complex mixture of chemical dyes in pH indicator solutions – or the glass membrane electrode in the pH meter.

The chemical or electrical response then has to be converted into a signal that we can observe, usually with our eyes. With litmus, this is easy. A colour change is observed, because of the change in the absorbance of visible light by the chemical itself, which is immediately detected by our eyes in a lightened room. In the case of the pH meter, the electrical response (a voltage change) has to be converted, i.e. *transduced* (= led through), into an observable response – movement of a meter needle or a digital display. The part of the device which carries out this conversion is called a *transducer*.

We can divide sensors into three types, namely (a) physical sensors for measuring distance, mass, temperature, pressure, etc. (which will not concern us here), (b) chemical sensors which measure chemical substances by chemical or physical responses, and (c) biosensors which measure chemical substances by using a biological sensing element. All of these devices have to be connected to a transducer of some sort, so that a visibly observable response occurs. Chemical sensors and biosensors are generally concerned with sensing and measuring particular chemicals which may or may not be biological themselves. We shall usually refer to such a material as the *substrate*, although the more general term *analyte* is sometimes used. Figure 1.1 shows schematically the general arrangement of a sensor.

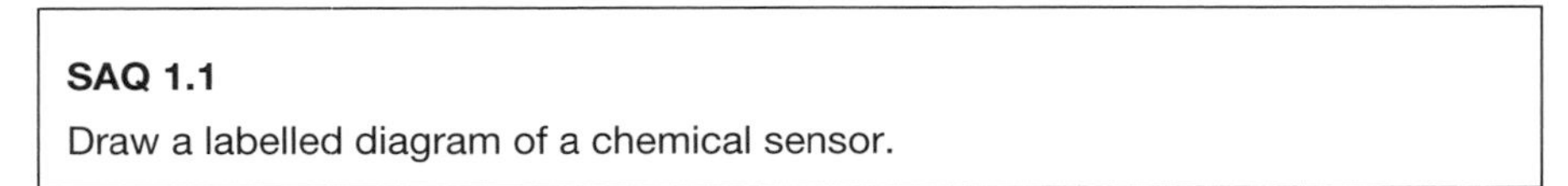

SAQ 1.1

Draw a labelled diagram of a chemical sensor.

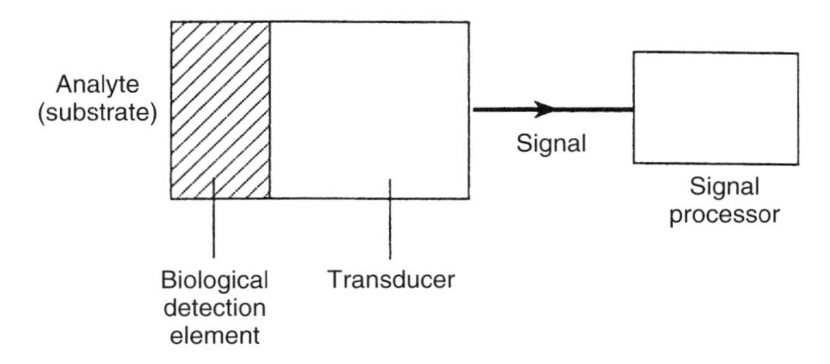

Figure 1.1 Schematic layout of a (bio)sensor. From Eggins, B. R., *Biosensors: An Introduction*, Copyright 1996. © John Wiley & Sons Limited. Reproduced with permission.

1.1.2 The Nose as a Sensor

One might consider the ears, eyes and fingers to be physical sensors as they detect physical sensations of sound, light and heat, etc., respectively. What we detect with the nose – smells – are in fact small quantities of chemicals. The nose is an extremely sensitive and selective instrument which is very difficult to emulate artificially. It can distinguish between many different chemical substances qualitatively and can give a general idea of 'quantity' down to very low detection limits. The chemicals to be detected pass through the olfactory membrane to the olfactory bulbs, which contain biological receptors that sense the substrate. The response is an electrical signal which is transmitted to the brain via the olfactory nerves. The brain then transduces this response into the sensation we know as smell. The tongue operates in a similar way.

Figure 1.2 shows a schematic diagram of the nasal olfactory system, illustrating the comparison with our generalized sensor. The nostrils collect the 'smell sample', which is then sensed by the olfactory membrane, i.e. the sensing element. The responses of the olfactory receptors are then converted by the olfactory nerve cell, which is the equivalent of the transducer, into electrical

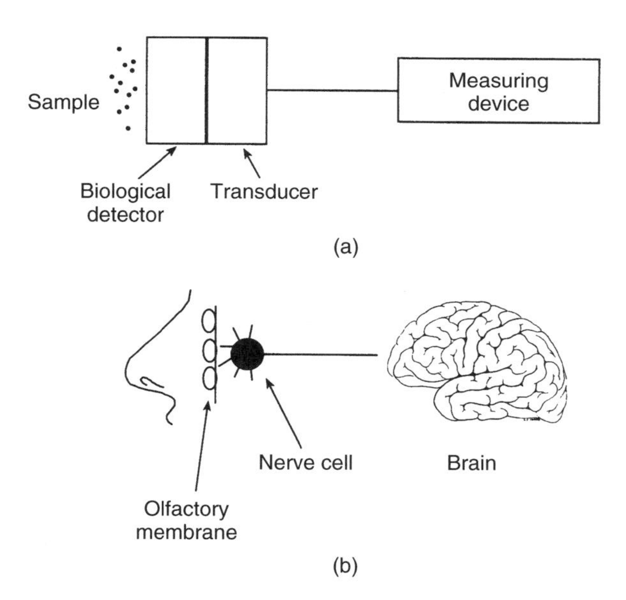

Figure 1.2 (a) Schematic of a sensor, showing the component parts, i.e. analyte, recognition element, transducer, actuator and measuring device. (b) Analogy with the nose as a sensor (actually a biosensor), in which the olfactory membrane is the biological recognition element, the nerve cell is the transducer, the nerve fibre is the actuator and the brain is the measuring element. From Eggins, B. R., *Biosensors: An Introduction*, Copyright 1996. © John Wiley & Sons Limited. Reproduced with permission.

signals which pass along the nerve fibre to the brain for interpretation. Thus, the brain acts as a microprocessor, turning the signal into a sensation which we call smell.

1.2 Sensors and Biosensors – Definitions

There are sometimes differences of usage for the terms *sensors, transducers, biosensors and actuators*, so it is necessary for us to define how they will be used in this book:

- We will use the term *sensor* to describe the whole device, following the Oxford English Dictionary definition, i.e. *a sensor is a device that detects or measures a physical property and records, indicates or otherwise responds to it.* (This is in contrast to the definition in Chambers Dictionary, quoted by Usher and Keating in their book 'Sensors and Transducers'.)
- We will define a *transducer* as *a device that converts an observed change (physical or chemical) into a measurable signal.* In chemical sensors, the latter is usually an electronic signal whose magnitude is proportional to the concentration of a specific chemical or set of chemicals.
- The term *actuator* i.e. *put into action*, is sometimes encountered. This is the part of the device which produces the display.

We can think of three types of sensor, i.e. physical, chemical and biosensors.

DQ 1.2

Distinguish between chemical sensors, physical sensors and biosensors.

Answer

Physical sensors are concerned with measuring physical quantities such as length, weight, temperature, pressure, and electricity – for their own sakes. This present book is not concerned with these as such except that the response of a sensor is usually in the form of a physical response. The book by Usher and Keating (see the Bibliography) is actually entirely concerned with physical sensors.

A chemical sensor is defined in R. W. Catterall's book (see the Bibliography) as **a device which responds to a particular analyte in a selective way through a chemical reaction and can be used for the qualitative or quantitative determination of the analyte.** *Such a sensor is concerned with detecting and measuring a specific chemical substance or set of chemicals.*

Biosensors are really a sub-set of chemical sensors, but are often treated as a topic in their own right. A biosensor can be defined as **a**

device incorporating a biological sensing element connected to a transducer. *The analyte that this sensor detects and measures may be purely chemical (even inorganic), although biological components may be the target analyte. The key difference is that the recognition element is biological in nature.*

SAQ 1.2

Distinguish between physical sensors, chemical sensors and biosensors.

1.3 Aspects of Sensors

1.3.1 Recognition Elements

Recognition elements are the key component of any sensor device. They impart the selectivity that enables the sensor to respond selectively to a particular analyte or group of analytes, thus avoiding interferences from other substances. Methods of analysis for specific ions have been available for a long time using ion-selective electrodes, which usually contain a membrane selective for the analyte of choice. In biosensors, the most common recognition element is an enzyme. Others include antibodies, nucleic acids and receptors.

1.3.2 Transducers – the Detector Device

Analytical methods in chemistry have mainly been based on photometric transducers, as in spectroscopic and colorimetric methods. However, most sensors have been developed around electrochemical transducers, because of simplicity of construction and cost. While electrons drive microprocessors, the directness of an electrical device will tend to have maximum appeal. However, with the rapid development of photon-driven devices through the use of optical fibres, it could well be that electrical appliances will soon become obsolete – starting with the telephone. In addition, the use of micro-mass-controlled devices, based mainly on piezo-electric crystals, may become competitive in the near future.

Transducers can be subdivided into the following four main types.

1.3.2.1 Electrochemical Transducers

(i) *Potentiometric*. These involve the measurement of the emf (potential) of a cell at zero current. The emf is proportional to the logarithm of the concentration of the substance being determined.

(ii) *Voltammetric*. An increasing (decreasing) potential is applied to the cell until oxidation (reduction) of the substance to be analysed occurs and there is a sharp rise (fall) in the current to give a peak current. The height of the peak current is directly proportional to the concentration of the electroactive material. If the appropriate oxidation (reduction) potential is known, one may step the potential directly to that value and observe the current. This mode is known as *amperometric*.

(iii) *Conductometric*. Most reactions involve a change in the composition of the solution. This will normally result in a change in the electrical conductivity of the solution, which can be measured electrically.

(iv) *FET-based sensors*. Miniaturization can sometimes be achieved by constructing one of the above types of electrochemical transducers on a silicon-chip-based field-effect transistor. This method has mainly been used with potentiometric sensors, but could also be used with voltammetric or conductometric sensors.

1.3.2.2 Optical Transducers

These have taken a new lease of life with the development of fibre optics, thus allowing greater flexibility and miniaturization. The techniques used include absorption spectroscopy, fluorescence spectroscopy, luminescence spectroscopy, internal reflection spectroscopy, surface plasmon spectroscopy and light scattering.

1.3.2.3 Piezo-electric Devices

These devices involve the generation of electric currents from a vibrating crystal. The frequency of vibration is affected by the mass of material adsorbed on its surface, which could be related to changes in a reaction. *Surface acoustic wave* devices are a related system.

1.3.2.4 Thermal Sensors

All chemical and biochemical processes involve the production or absorption of heat. This heat can be measured by sensitive thermistors and hence be related to the amount of substance to be analysed.

Further details of all of these types of transducers are given below in Chapters 2 and 7.

SAQ 1.3

Summarize the main features of sensors.

1.3.3 Methods of Immobilization

The selective element must be connected to the transducer. This presents particular problems if the former is biological in nature. Several classes of methods of connection have evolved, as follows:

- The simplest method is *adsorption* on to a surface.
- *Microencapsulation* is the term used for trapping between membranes – one of the earliest methods to be employed.
- *Entrapment*, where the selective element is trapped in a matrix of a gel, paste or polymer – this is a very popular method.
- *Covalent attachment*, where covalent chemical bonds are formed between the selective component and the transducer.
- *Cross-linking*, where a bifunctional agent is used to bond chemically the transducer to the selective component – this is often used in conjunction with other methods, such as adsorption or microencapsulation.

Further details about these various procedures are presented below in Section 3.6.

1.3.4 Performance Factors

(i) *Selectivity*. This is the most important characteristic of sensors – the ability to discriminate between different substances. Such behaviour is principally a function of the selective component, although sometimes the operation of the transducer contributes to the selectivity.

(ii) *Sensitivity range*. This usually needs to be sub-millimolar, but in special cases can go down to the femtomolar (10^{-15} M) range.

(iii) *Accuracy*. This needs to be better than ±5%.

(iv) *Nature of solution*. Conditions such as pH, temperature and ionic strength must be considered.

(v) *Response time*. This is usually much longer (30 s or more) with biosensors than with chemical sensors.

(vi) *Recovery time*. This is the time that elapses before the sensor is ready to analyse the next sample – it must not be more than a few minutes.

(vii) The *working lifetime* is usually determined by the stability of the selective material. For biological materials this can be a short as a few days, although it is often several months or more.

Further details of the above are given later in Chapter 4.

1.3.5 Areas of Application

1.3.5.1 Health Care

Health care is the main area of application of biosensors and chemical sensors (*chemisensors*). Measurements of blood, gases, ions and metabolites are regularly needed to show a patient's metabolic state – especially for those in hospital, and

Table 1.1 Common assays that are required in diagnostic medicine

Analyte	Method of assay
Glucose	Amperometric biosensor
Urea	Potentiometric biosensor
Lactate	Amperometric biosensor
Hepatitis B	Chemiluminescent immunoassay
Candida albicans	Piezo-electric immunoassay
Cholesterol	Amperometric biosensor
Penicillins	Potentiometric biosensor
Sodium	Glass ion-selective electrode
Potassium	Ion-exchange-selective electrode
Calcium	Ionophore ion-selective electrode
Oxygen	Fluorescent quenching sensor
pH	Glass ion-selective electrode

even more so if they are in intensive care. Many of these substrates have been determined by samples of urine and blood being taken away to a medical analytical laboratory for classical analysis, which may not be complete for hours or even days. The use of on-the-spot sensors and biosensors enable results to be obtained in minutes at most. The latest ExacTech® glucose sensor gives a reading in 12 s.

This would obviate the need for *en suite* analytical units with specialist medical laboratory scientists. A trained nurse would be competent to carry out sensor tests at the bedside. Modern 'smart' sensors, based on field-effect transistors (FETs), may combine several measurements in one sensor unit. This particularly applies in the case of ion sensors for sodium, potassium, calcium and pH. Attempts are also being made to make combination biosensors, e.g. for glucose, lactate and urea. Table 1.1 shows a list of common assays that are routinely needed for diagnostic work with patients.

A potential 'dream application' is to have an implanted sensor for continuous monitoring of a metabolite. This might then be linked via a microprocessor to a controlled drug-delivery system (e.g. an iontophoretic system) through the skin. Such a device would be particularly attractive for chronic conditions such as diabetes. The blood glucose sensor would be monitored continuously and, as the glucose level reached a certain value, insulin would be released into the patient's blood stream automatically. This type of system is sometimes referred to as an *artificial pancreas*. The latter would be far more beneficial for the patient than the present system of discrete blood glucose analyses which involve pricking a thumb every time, followed by injection of large doses of insulin every few hours.

1.3.5.2 Control of Industrial Processes

Sensors are used in various aspects of fermentation processes in three different ways, i.e. (i) off-line in a laboratory, (ii) off-line, but close to the operation site, and (iii) on-line in real time. At present, the main real-time monitoring is confined to such measurements as temperature and pH, plus carbon dioxide and oxygen measurements. However, biosensors which monitor a range of direct reactants and products are available, such as those for sugars, yeasts, malts, alcohols, phenolic compounds, and perhaps, undesirable by-products. Such monitoring could result in improved product quality, increased product yields, checks on tolerance of variations in quality of raw material, optimized energy efficiency, i.e. improved plant automation, and less reliance on human judgement. In general, there is a wide range of applications in the food and beverage industry.

1.3.5.3 Environmental Monitoring

There is an enormous range of potential analytes in air, water, soils, and other environmental situations. Such measurements in water include biochemical oxygen demand (BOD), acidity, salinity, nitrate, phosphate, calcium and fluoride, while pesticides, fertilizers and both industrial and domestic wastes require extensive analyses. A current concern is for endocrine disruptors that can be active at very low levels of concentration (ng l^{-1}), due to a wide range of oestrogens and oestrogenic mimics. Continuous real-time monitoring is required for some substances, and occasional random monitoring for others. In addition to the obvious pollution applications, farming, gardening, veterinary science and mining are all areas where sensors are needed for environmental monitoring.

SAQ 1.4

Write down an example of a sensor used in (a) health care, (b) industrial control, and (c) environmental monitoring.

Summary

This chapter discusses what sensors are, defining physical, chemical and biosensors. The various parts of a sensor are described, including (the analyte), recognition element, transducer, actuator, and measuring device. The biosensor is compared to the human nose. An introduction is given to various aspects of sensors, including types of recognition elements, transducers, connection methods, and performance factors. Three main areas of application are described.

Chapter 2

Transduction Elements

Learning Objectives

- To set up an electrochemical cell.
- To appreciate the reason for references electrodes.
- To know the main types of reference electrodes.
- To know the relationship between electrode potential and analyte concentration.
- To set up a concentration cell for the measurement of ionic concentrations.
- To be able to describe the main types of ion-selective electrodes.
- To understand the principles of interference with ion-selective electrodes.
- To be able to describe the three ways in which current is conveyed through an electrolyte.
- To be able to draw and explain diagrams showing current–voltage curves (voltammograms).
- To know the relationship between voltammetric (or amperometric) current and analyte concentration.
- To explain why various modes of electrode modification may improve selectivity.
- To be able to draw a diagram of an insulated-gate field-effect transistor (IGFET).
- To describe the application to CHEMFETs, ISFETs, and ENFETs.
- To explain the purpose of thin-film electrodes and thick-film (screen-printed) electrodes.
- To describe the advantages and disadvantages of microelectrodes.
- To state the Beer–Lambert law.
- To know the range of optical procedures and responses suitable for sensors.

- To know the mechanisms for the production of fluorescence and its quenching.
- To know the relationships between response and analyte concentration.
- To describe chemiluminescence and bioluminescence.
- To explain the operation of optical fibre waveguides.
- To know how optical sensing agents are immobilized to form optodes.

2.1 Electrochemical Transducers – Introduction

There are three basic electrochemical processes that are useful in transducers for sensor applications:

(i) *Potentiometry*, the measurement of a cell potential at zero current.

(ii) *Voltammetry* (amperometry), in which an oxidizing (or reducing) potential is applied between the cell electrodes and the cell current is measured.

(iii) *Conductometry*, where the conductance (reciprocal of resistance) of the cell is measured by an alternating current bridge method.

2.2 Potentiometry and Ion-Selective Electrodes: The Nernst Equation

2.2.1 Cells and Electrodes

When a piece of metal (such as silver) is placed in a solution containing ions (such as silver ions), there is a charge separation across the boundary between the metal and the solution (Figure 2.1). This sets up what we can call an *electron*

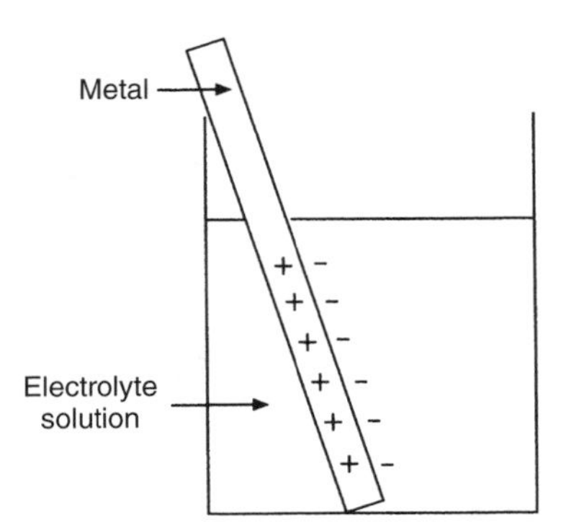

Figure 2.1 A metal electrode dipped into an electrolyte solution – one half-cell. From Eggins, B. R., *Biosensors: An Introduction*, Copyright 1996. © John Wiley & Sons Limited. Reproduced with permission.

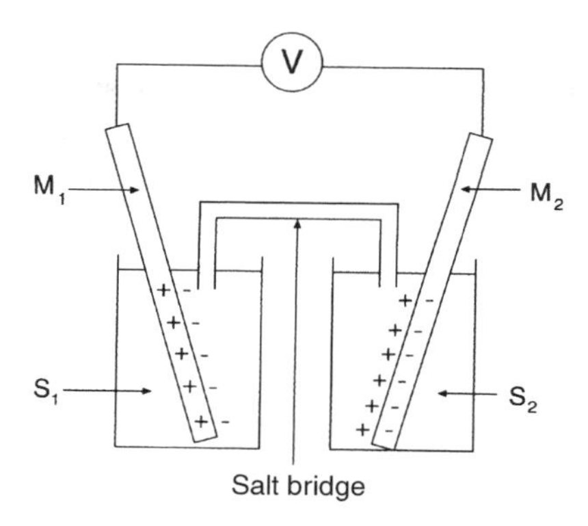

Figure 2.2 Two half-cell electrodes combined, making a complete cell. From Eggins, B. R., *Biosensors: An Introduction*, Copyright 1996. © John Wiley & Sons Limited. Reproduced with permission.

pressure, usually termed a *potential*. It cannot be measured directly, and requires two such electrode–electrolyte combinations. Each of these is called a *half-cell*. Such a combination is called an electrochemical *cell* (Figure 2.2).

The two half-cells must be connected internally by means of an electrically conducting bridge or membrane. Then, the two electrodes are connected externally by a potential measuring device, such as a digital voltmeter (DVM). This has a very high internal impedance ($\sim 10^{-12}$ Ω), such that very little current will flow through it. The electrical circuit is now complete and the emf of the cell can be measured. This value is the difference between the electrode potentials of the two half-cells. Its magnitude depends on a number of factors, i.e. (i) the nature of the electrodes, (ii) the nature and concentrations of the solutions in each half-cell, and (iii) the liquid junction potential across the membrane (or salt bridge).

SAQ 2.1

Why are two half-cells needed in order to be able to measure a cell emf?

The practical Daniell cell is a good example of an electrochemical cell (see Figure 2.3). Such a cell involves copper and zinc electrodes in solutions of copper(II) and zinc(II) sulfates, with a porous pot for the bridge. To keep matters simple, we shall assume that the concentrations of the electrolytes are both 1 M. This cell has been used as a practical battery and has an emf of 1.10 V.

We can consider the half-cell reactions as follows:

$$Cu^{2+} + 2e^- = Cu \quad (2.1)$$

$$Zn^{2+} + 2e^- = Zn \quad (2.2)$$

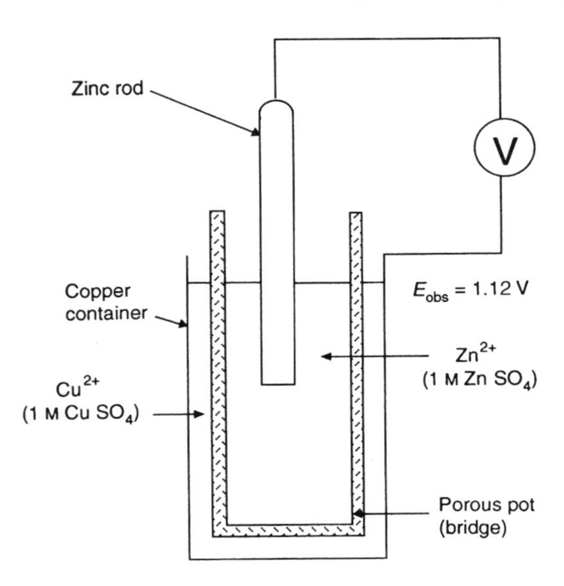

Figure 2.3 Schematic of the Daniell cell. From Eggins, B. R., *Biosensors: An Introduction*, Copyright 1996. © John Wiley & Sons Limited. Reproduced with permission.

If we subtract equation (2.2) from equation (2.1), we obtain the complete cell reaction, as follows:

$$Cu^{2+} + Zn = Cu + Zn^{2+}$$

The Gibbs free energy for this reaction is negative, showing that the reaction will proceed spontaneously in the direction from left to right. The reaction can easily be carried out directly in a test tube, by the addition of copper(II) sulfate solution to pieces of zinc. The white zinc metal quickly becomes covered with a dark brown coating of copper metal and the blue colour of the copper(II) sulfate fades as it is replaced by colourless zinc sulfate. The Gibbs free energy is simply related to the emf of the cell by the following expression:

$$\Delta G = -nFE \tag{2.3}$$

where n is the number of electrons transferred (in this case, $n = 2$), F is the Faraday constant ($= 96\,487$ C mol^{-1}), and E is the emf of the cell (if we assume that the liquid junction potential is zero). Thus, if ΔG is negative, then E is positive.

We may now ask what the ΔG values are for reactions (2.1) and (2.2) separately. If we could determine ΔG_{Cu} and ΔG_{Zn}, we could then find E_{Cu} and E_{Zn} separately. A simple separation is not possible, however, so we must take another approach. Consider the first element in the Periodic Table, *hydrogen*. This is not a metal, but it can be oxidized to hydrogen ions, H^+, by the removal of an electron, as follows:

$$H - e^- = H^+$$

which is more usually written as:

$$H^+ + e^- = 1/2\ H_2$$

The Gibbs free energy (ΔG) for this reaction is defined as zero for the standard state (which is when the concentration of H^+ is 1 M, the partial pressure of hydrogen gas is 1 atm and the temperature is 298 K (25°C)). For any standard state, the Gibbs free energy is designated as ΔG^0. The standard electrode potential for hydrogen is therefore:

$$E_H{}^0 = 0$$

We can set up a practical half-cell hydrogen electrode, which can be combined with any other half-cell, as shown in Figure 2.4. We can show the half-cell reactions as before:

$$Cu^{2+} + 2e^- = Cu \quad (E_{Cu}{}^0) \tag{2.4}$$

$$2H^+ + 2e^- = H_2 \quad (E_H{}^0) \tag{2.5}$$

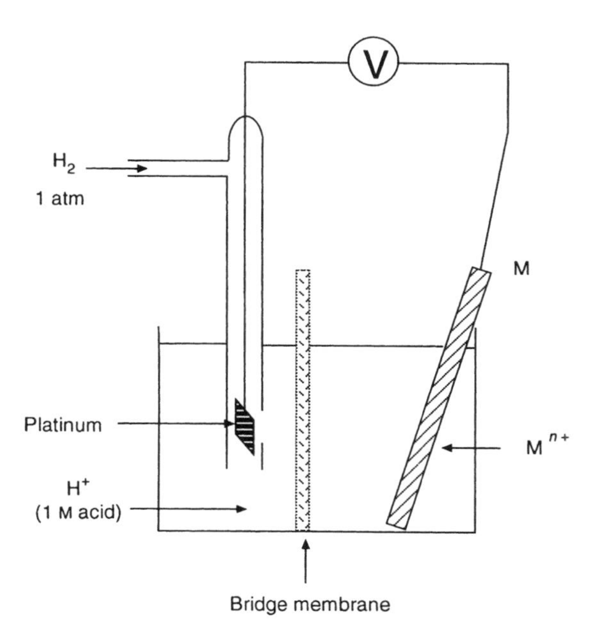

Figure 2.4 A hydrogen electrode connected with another half-cell. From Eggins, B. R., *Biosensors: An Introduction*, Copyright 1996. © John Wiley & Sons Limited. Reproduced with permission.

By subtracting equation (2.5) from (2.4), we obtain the following:

$$Cu^{2+} + H_2 = Cu + 2H^+$$

Thus:

$$E_{cell} = E_{Cu}{}^0 - E_H{}^0 = +0.34 \text{ V}$$

and therefore:

$$E_{Cu}{}^0 = +0.34 \text{ V}$$

Hence, a scale of E values can be set up for a whole range of half-cell electrodes against the *standard hydrogen electrode* (SHE).

For the other half of the Daniell cell, i.e. the zinc electrode, we have the following:

$$Zn^{2+} + 2e^- = Zn \quad (E_{Zn}{}^0) \tag{2.6}$$

$$2H^+ + 2e^- = H_2 \quad (E_H{}^0) \tag{2.7}$$

Thus, by subtracting equation (2.7) from (2.6), we obtain:

$$Zn^{2+} + H_2 = Zn + 2H^+$$

Thus:

$$E_{cell} = E_{Zn}{}^0 - E_H{}^0 = -0.76 \text{ V}$$

and therefore:

$$E_{Zn}{}^0 = -0.76 \text{ V}$$

Combining the half-cells for copper and zinc gives the cell emf for the Daniell cell as follows:

$$E_{cell} = +0.34 - (-0.76) = 1.10 \text{ V}$$

SAQ 2.2

What is the cell reaction of a galvanic cell represented by the following notation:

$$Pt|Fe^{3+}, Fe^{2+}||Cu^{2+}|Cu?$$

2.2.2 Reference Electrodes

The standard hydrogen electrode is a *reference electrode* (RE) to which other electrodes may be referred. While it is not difficult to set up an SHE in the laboratory, it is not very convenient for routine measurements as such an electrode involves flowing hydrogen gas, which is potentially explosive. Other secondary reference electrodes are therefore used in practice – these are easy to set up, are non-polarizable and give reproducible electrode potentials which have low coefficients of variation with temperature.

Many varieties of these electrodes have been devised, but two are in common use and are easy to set up and also available commercially.

2.2.2.1 The Silver–Silver Chloride Electrode

Silver chloride has the advantage of being sparingly soluble in water. The half-cell reaction is as follows:

$$AgCl + e^- = Ag + Cl^- \quad (E^0 = +0.22\ V)$$

This electrode consists of a silver wire coated with silver chloride dipping into a solution of sodium chloride.

2.2.2.2 The Saturated-Calomel Electrode

'Calomel' is the old-fashioned name for mercurous chloride (Hg_2Cl_2). The half-cell reaction is similar to that shown above for the silver–silver chloride electrode:

$$Hg_2Cl_2 + 2e^- = 2Hg + 2Cl^- \quad (E^0 = 0.24\ V)$$

This electrode consists of a mercury pool in contact with a paste made by mixing mercury(I) chloride powder and saturated potassium chloride solution, with

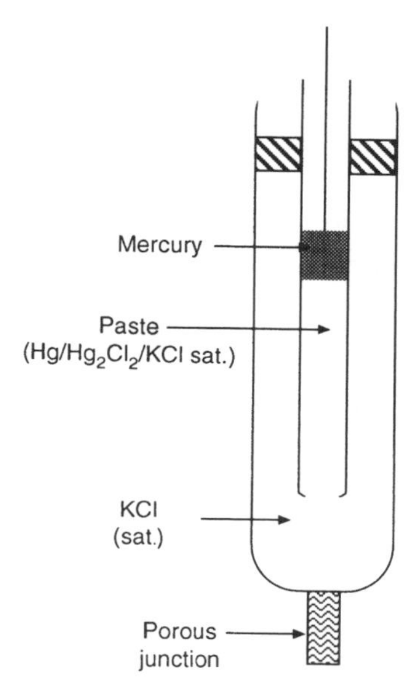

Figure 2.5 Schematic of a saturated-calomel electrode. From Eggins, B. R., *Biosensors: An Introduction*, Copyright 1996. © John Wiley & Sons Limited. Reproduced with permission.

the whole being in contact with a saturated solution of potassium chloride (see Figure 2.5). The advantage of the latter is that it can be easily obtained by simply shaking potassium chloride with water until no more dissolves. Thus, one has a solution of exact and reproducible concentration without the need for measurements of weights or volumes.

These electrodes are suitable for most purposes involving aqueous solutions. Other types are available for use in non-aqueous solutions or if chloride ions must be absent. One usually measures the potential difference between an indicator electrode and the reference electrode to give the cell emf, as follows:

$$E_{\text{cell}} = \Delta E = E_{\text{ind}} - E_{\text{REF}}$$

DQ 2.1

Discuss the meaning of the term *potentiometry*.

Answer

Potentiometry is an analytical technique which involves the measurement of the emf of a cell at equilibrium, either to directly determine the concentration (activity) of an ion by using the Nernst equation (direct potentiometry), or to detect the end-point in a titration (potentiometric titration).

SAQ 2.3

What are the characteristics needed for a reference electrode?

2.2.3 Quantitative Relationships: The Nernst Equation

So far, we have only considered electrode potentials at one concentration of oxidized species (Ox) or reduced species (R), usually 1 M. Now we must consider the effect of different concentrations on the electrode potential. This is of fundamental importance for analytical applications of potentiometry. The basic Nernst equation is a logarithmic relationship derived from fundamental thermodynamic equations such as the following:

$$\Delta G = RT \ln K$$

So, for the half-cell reaction, we have:

$$\text{Ox} + n\text{e}^- = \text{R}$$

The Nernst equation is as follows:

$$E = E^0 + \frac{RT}{nF} \ln \left(\frac{a_{\text{Ox}}}{a_{\text{R}}} \right)$$

where a_{Ox} and a_R are activities, i.e. ideal thermodynamic concentrations, which for dilute solutions can be taken to be the same as (conventional) concentrations.

It is usually more useful to express concentrations in powers of ten and therefore to use logarithms to base 10 rather than *natural* logarithms to base e. The Nernst equation then becomes:

$$E = E^0 + 2.303\frac{RT}{nF}\log_{10}\left(\frac{[\mathrm{Ox}]}{[\mathrm{R}]}\right)$$

It should be noted that this equation has the same form as the Henderson–Hasselbach equation for the pH of mixtures of acids and bases:

$$\mathrm{pH} = \mathrm{p}K_a + \log_{10}\left(\frac{[\mathrm{A}^-]}{[\mathrm{HA}]}\right)$$

Logarithms

Logarithms are less familiar to students now that 'log tables' are no longer an essential tool for calculations – multiplications, divisions, powers and roots. However, the idea of the logarithm is still essential for much scientific work. This is a way of expressing a value in terms of a power. Thus, we commonly say that $2 \times 2 = 2^2$ = 'two squared' = 4. Similarly, $2 \times 2 \times 2 = 2^3$ = 'two cubed' = 8, and so on. The power to which 2 is raised, i.e. 2 or 3, is called the 'logarithm' (to base 2) of the numbers 4 and 8, respectively. When we measure very large numbers such as the speed of light, which is 300 000 000 m s^{-1}, it is often more convenient to express them as powers of ten, i.e. 3×10^8 m s^{-1}. Similarly, very small numbers such as concentrations of very dilute solutions, e.g. 0.000 000 01 M, i.e. 1/100 000 000 M, can better be expressed as 1×10^{-8} M. In these examples, the powers or indices +8 and −8 are the logarithms (to base 10). In scientific work, it happens that a special base for logarithms, called 'e', is used. These are known as 'natural logarithms'. The value of e is 2.718. . . . This need not worry us as logarithms to base e are simply related to logarithms to base 10 by the value 2.303. Thus, the natural logarithm of x (called $\ln x$ or $\log_e x$) is just $2.303 \log_{10} x$. In fact, with a calculator this conversion is rarely needed as scientific calculators give values for both natural logarithm (to base e) and common logarithms (to base 10). To get back to a number from a logarithm, one needs the index, or anti-logarithm as it used to be called. The scientific calculator also gives this.

A calculator key is labelled:

10^x
LOG

The main function gives the common logarithm (to base 10) of a number. The inverse (or second) function gives the anti-logarithm, 10^x.

The adjacent key will be labelled:

e^x
LN

This similarly gives the natural logarithm (ln) of x, while the inverse gives e^x. A little practice is needed with these concepts if they are unfamiliar to you, so that a facility may be acquired with the manipulation of concentrations in terms of powers of ten.

Measurements of acidity (hydrogen ion concentrations) are normally expressed as pH values, where pH *is* a logarithm. Thus, for an acid concentration of 10^{-3} M, we take the positive value of the negative index, i.e. 3, so the pH value is 3. In general:

$$\mathrm{pH}(x) = -\log_{10}(x)$$

A slightly more complex, but very common example relates to the pH of 1 M acetic acid (a weak acid). The hydrogen ion concentration is 1.8×10^{-5}, so pH $= -\log_{10}(1.8 \times 10^{-5}) = 4.745$. Thus, $10^{-4.745} = \log(1.8 \times 10^{-5})$.

If we put in the values of the constants $R = 8.314$ J K^{-1} mol^{-1} and $F =$ 96 480 C mol^{-1}, plus a value for T, we obtain the following:

$$RT/F = 0.0257\ \mathrm{V}, \quad \text{at } T = 298\ \mathrm{K}\ (25^\circ\mathrm{C})$$

$$RT/F = 0.0252\ \mathrm{V}, \quad \text{at } T = 293\ \mathrm{K}\ (20^\circ\mathrm{C})$$

$$2.303(RT/F) = 0.0591, \quad \text{at } T = 298\ \mathrm{K}$$

$$2.303(RT/F) = 0.0580, \quad \text{at } T = 293\ \mathrm{K}$$

It is often useful to approximate these latter values to 0.06, thus giving a simplified form of the Nernst equation, as follows:

$$E = E^0 + 0.06 \log\left(\frac{[\mathrm{Ox}]}{[\mathrm{R}]}\right)$$

The reduced species, R, is often a metal, in which case it has a constant concentration (activity) of 1, so the equation simplifies further to the following:

$$E = E^0 + 0.06 \log [\mathrm{Ox}]$$

We can generalize this for practical situations in which E^0 and $2.303(RT/F)$ may not be known or may differ from the theoretical values:

$$E = K + S \log [\mathrm{Ox}]$$

This equation is a very useful practical form of the Nernst equation. As we shall see later from experimental data, we can plot a graph of E against $\log [\mathrm{Ox}]$, which would normally give a straight line of slope S, with an intercept of K. Then, the experimental values of S and K can be compared with the theoretical values, i.e. $S = 2.303(RT/F)$ and $K = E^0$.

If we now incorporate the reference electrode potential (E_{REF}) and the liquid junction potential (E_{lj}), as shown in Figure 2.6, we have the following:

$$E_{\mathrm{cell}} = E'_{\mathrm{M/M}^{n+}} - E_{\mathrm{REF}} - E_{\mathrm{lj}}$$

$$= E'_{\mathrm{M/M}^{n+}} = E'^{0} + S \log [\mathrm{M}^{n+}]$$

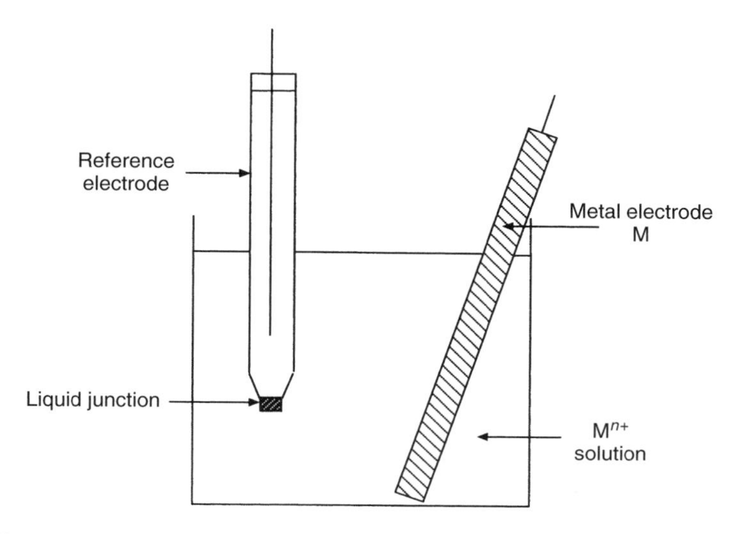

Figure 2.6 A reference electrode combined with another half-cell. From Eggins, B. R., *Biosensors: An Introduction*, Copyright 1996. © John Wiley & Sons Limited. Reproduced with permission.

Therefore:

$$E_{\text{cell}} = (E'^{0} - E_{\text{REF}} - E_{\text{lj}}) + S \log [\text{M}^{n+}]$$

and hence:

$$E_{\text{cell}} = K + S \log [\text{M}^{+}]$$

with:

$$K = (E'^{0} - E_{\text{REF}} - E_{\text{lj}})$$

SAQ 2.4

Show how the Nernst equation may be simplified to the following form:

$$E = K + S \log C$$

What are K and S?

SAQ 2.5

A galvanic cell consisting of an SHE (as the left-hand electrode) and a rod of zinc dipping into a solution of zinc ions at 298 K gave a measured emf of −0.789 V. What is the activity of the zinc ions?

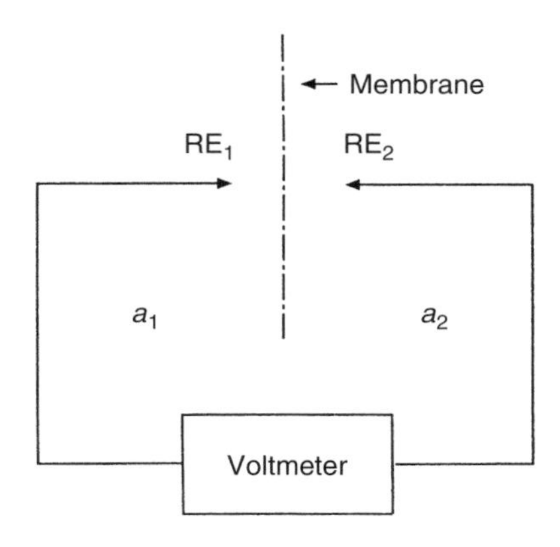

Figure 2.7 Schematic of a concentration cell; RE_1 and RE_2 represent reference electrodes. From Eggins, B. R., *Biosensors: An Introduction*, Copyright 1996. © John Wiley & Sons Limited. Reproduced with permission.

If instead of a reference electrode, we incorporate a similar half-cell with the same redox couple but with a different concentration of Ox, as shown in Figure 2.7, we can set up the two half-cell reactions as follows:

$$E_1 = E^0 + S \log [\mathrm{Ox}]_1$$

$$E_2 = E^0 + S \log [\mathrm{Ox}]_2$$

By subtracting the second equation from the first, we obtain:

$$\Delta E = E_1 - E_2 = S \log ([\mathrm{Ox}]_1/[\mathrm{Ox}]_2)$$

Now, if $[\mathrm{Ox}]_2$ is kept constant (perhaps at a reference concentration), we then obtain:

$$\Delta E = \text{constant} + S \log [\mathrm{Ox}]_1$$

where the constant is $-S \log [\mathrm{Ox}]_2$.

This result is made use of in a practical way with most forms of ion-selective electrodes. The general practical arrangement is shown in Figure 2.7.

On the left in this figure is the test solution to be determined, into which is dipped a reference electrode. The ion-selective membrane is in the middle dividing the test solution from the standard solution on the right, which consists of a fixed concentration of the ions being measured. Into this is placed a second reference electrode. The two reference electrodes are connected through a high-impedance voltmeter – usually a digital voltmeter. The electrode system is then usually calibrated with standard solutions in one of a number of ways, as described below.

The observed voltage is the difference between the two half-cell electrodes, which are identical, except that the concentrations of the ion being determined differ in each half of the cell. We can write the voltage (emf) of the cell as follows:

$$E = E_{RE_1} + E_{RE_2} + E_{lj} - S \log a_2 + S \log a_1$$

and so:

$$E = K + S \log a_1$$

where:

$$K = (E_{RE_1} + E_{RE_2} + E_{lj} - S \log a_2)$$

and a_1 and a_2 are the activities of the test and reference (standard) solutions, respectively.

DQ 2.2

What is an ISE?

Answer

An ISE is an **ion-selective electrode**, *designed to respond to one particular ion more than others. This is a potentiometric device, i.e. the potential of the electrode, measured against an appropriate reference electrode, is proportional to the logarithm of the activity (or concentration) of the ion being tested. Such a device usually responds rapidly, with a linear range of about 1 to* 10^{-6} *M for most ISEs. It operates on the principle of a concentration cell, in that it contains a selective membrane which develops a potential if there is a concentration difference across the membrane of the ion being tested.*

2.2.4 Practical Aspects of Ion-Selective Electrodes

In order to obtain consistent, reproducible results with the lowest detection limits, certain precautions have to be observed. Sample standardization does not involve extensive pre-treatment. Usually the addition of a special buffer is sufficient. The following factors may need to be observed:

(i) The ionic strength needs to be kept constant from one sample to the next. This can simply be done by adding a fairly high, constant concentration of an indifferent electrolyte, i.e. one that does not interfere in any way, to each sample and each standard.

(ii) The pH may need to be controlled at a certain level. This is more important with some ionic samples than others, e.g. fluorides.

(iii) It may be possible, and desirable, to add components that minimize or eliminate interfering ions.

Appropriate mixtures to provide these properties are usually called ionic-strength adjusters (ISAs) or more fully, total-ionic-strength adjustment buffers (TISABs).

For example, with nitrate ISEs the ISA is commonly just sodium sulfate. This, although not strictly a pH buffer, keeps the pH well within the required 2 to 12 limit. However, it does not eliminate the considerable and important interferences from chloride and nitrite that might be present. Alternative ISAs of more complex composition will achieve this. Silver sulfate was originally used to precipitate out chloride ions, but this has now been replaced by a more complex but less expensive lead acetate mixture.

2.2.5 Measurement and Calibration

2.2.5.1 Calibration Graphs and Direct Reading

This is the most straightforward method. A series of standard solutions are made up with an added ISA and the potentials are measured. Then, a calibration graph is plotted of voltage against log (concentration). Deviations from linearity or a Nernstian slope do not matter. The sample is treated in the same way and its log (concentration) value is then read from the graph. New calibration graphs should be prepared regularly.

2.2.5.2 Standard Addition

The sample is prepared as before and its voltage is read. Then, a known amount of a standard of higher concentration, usually about 10 times the expected sample concentration, is added and a second voltage reading is taken. The data are then fitted to an equation, which should include a correction for dilution by the added standard.

If C_u is the unknown concentration in V_u ml of solution and C_s is the added standard concentration in V_s ml of solution, then we have:

$$E_1 = K + S \log C_u$$

and

$$E_2 = K + S \log (C_u V_u + C_s V_s)/(V_u + V_s)$$

Subtracting the second equation from the first, we obtain:

$$E = S \log \{C_u/[C_u V_u + C_s V_s)/(V_u + V_s)]\}$$

This can be rearranged to give the following:

$$C_u = C_s/\{10^{E/S}[1 + (V_u/V_s)] - V_u/V_s\}$$

Hence, C_u can be obtained.

2.2.5.3 Gran Plot

This is really an extension of the standard addition method, using multiple standard additions. The procedure is the same as in the standard addition method except that several additions are made (say, five or more). By using the above nomenclature, except that in this case the single-value C_s is replaced by a variable C_s, where the latter represents the increase in concentration in the sample solution produced by each addition, we have the following:

$$E = K + S \log (C_u + C_s)$$

and therefore:

$$E/S = K/S + \log (C_u + C_s)$$

By taking anti-logarithms, we obtain:

$$10^{E/S} = K'(C_u + C_s)$$

where $K' = 10^{K/S}$. A plot of $10^{E/S}$ against C_s (a Gran plot) is shown in Figure 2.8. The plot is a straight line, with a negative intercept of $-C_u$, as when $10^{E/S} = 0$, $C_u = -C_s$. This derivation does not show corrections for added volumes of standards.

SAQ 2.6

How is the emf of a cell related to the concentration of the analyte?

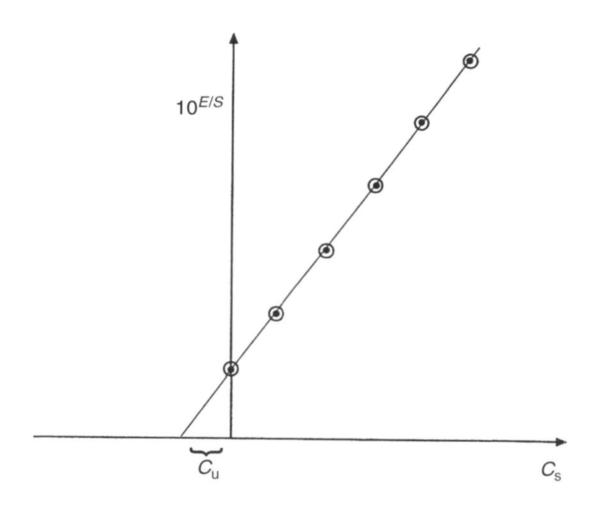

Figure 2.8 An example of a typical Gran plot. From Eggins, B. R., *Biosensors: An Introduction*, Copyright 1996. © John Wiley & Sons Limited. Reproduced with permission.

DQ 2.3

Discuss the relative merits of calibration graphs, single standard addition methods and multiple standard addition methods, with respect to accuracy and convenience.

Answer

A calibration graph can be used in most circumstances where accuracy depends on the number of data points. Any unknown can be determined whose concentration falls within the range of the graph (or even just outside of this range if it is linear). Such a graph can still be used if the slope is 'non-Nernstian' or indeed if there is some curvature present. However, this graph can take time to prepare, as it requires the use of between 5 and 10 standard solutions.

A single standard addition requires only one standard solution, which does, however, need to be reasonably matched to the expected value of the unknown. This approach assumes that there is a linear relationship between the emf and the logarithm of the concentration and the slope is 'Nernstian' (59 mV per decade). One could operate with a different slope, provided that the actual slope itself had been determined from a calibration graph. The calculation involved in this case is not very obvious.

The multiple standard addition method combines the technique of adding standards to the unknown solution, plus the use of multiple standards. This approach is sometimes more convenient than the calibration graph method, but as with the single standard addition technique one needs to know that a linear relationship applies, as well as a knowledge of the value of the slope. However, the corresponding graph is easy to plot and gives a result as a negative intercept without any further calculations.

SAQ 2.7

The following data were obtained for the calibration of a calcium ISE and an unknown sample S:

[Ca] (M)	E (mv)
1.00×10^{-4}	-2.0
5.00×10^{-4}	$+16.0$
1.00×10^{-3}	$+25.0$
5.00×10^{-3}	$+43.0$
1.00×10^{-2}	$+51.0$
[S]	$+33.0$

What is the concentration of calcium in the sample S?

DQ 2.4

Discuss the similarities and differences between the different types of ion-selective electrodes.

Answer

All ion-selective electrodes contain a selective membrane behind which is a standard solution containing the ion being tested and an internal reference electrode. The differences are in the types of membrane that are used. The glass membrane used in pH electrodes is very familiar. However, the other types all look the same from the outside. The solid-sensor type uses a single crystal or a pressed powder pellet, whereas the liquid ion-exchange type uses a liquid ion exchanger soaked into a porous pad, which dips into a reservoir containing the ion-exchange liquid.

2.3 Voltammetry and Amperometry

2.3.1 Linear-Sweep Voltammetry

The above two terms cover a range of techniques involving the application of a linearly varying potential between a working electrode and a reference electrode in an electrochemical cell containing a high concentration of an indifferent electrolyte to make the solution conduct – called the supporting electrolyte – and an oxidizable or reducible species – the electroactive species.

The current through the cell is monitored continuously. A graph is traced on a recorder of current against potential – this is known as a *voltammogram*. The most straightforward technique is called *linear-sweep voltammetry* (LSV). A typical voltammogram is shown in Figure 2.9.

At the start (point A) the current is very small. Between points A and B, it rises very slowly owing to the residual current (from impurities) and double-layer charging (where the electrode–solution interface acts as a capacitor). This is sometimes called the *background current*. At point B, the potential approaches the reduction potential of the oxidized species (Ox). The increasing potential causes electrons to transfer from the electrode to the Ox at an increasing rate, according to the following general reaction:

$$\text{Ox} + n\text{e}^- \longrightarrow \text{R}$$

The increasing rate of reduction causes the cell current to increase. It can be shown that the net cell current in this region is given by the algebraic sum of a cathodic (reduction) current (i_c) and an anodic current (i_a), as follows:

$$i_{\text{net}} = i_c + i_a$$

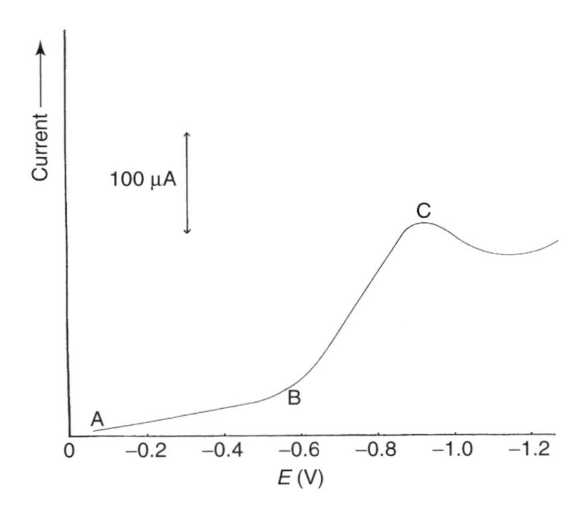

Figure 2.9 A linear sweep voltammogram. From Eggins, B. R., *Biosensors: An Introduction*, Copyright 1996. © John Wiley & Sons Limited. Reproduced with permission.

where:

$$i_c = nFAk_f^0 C_{Ox} \exp[-\alpha nF(E - E_{eq})/RT]$$

and

$$i_a = -nFAk_b^0 C_R \exp[(1 - \alpha)nF(E - E_{eq})/RT]$$

with E_{eq} being the equilibrium potential.

As E (the applied potential) increases, i_c increases and i_a decreases, thus causing the observed rise in the voltammetric wave. This rise does not continue indefinitely, however, as it is limited by the fact that the concentration of Ox becomes depleted by the reduction process, while the current is limited by the decreasing rate of diffusion of fresh Ox from the bulk of the solution. This results in a peak in the current at point C in Figure 2.9. This diffusion effect is shown qualitatively in Figure 2.10.

The value of the diffusion-limited current (i_d) is obtained from Fick's first law of diffusion, as follows:

$$\text{flux of material to electrode surface} = \mathrm{d}N/\mathrm{d}t = D\,\mathrm{d}C/\mathrm{d}x \tag{2.8}$$

thus giving:

$$i_d = nFAD\,\mathrm{d}C/\mathrm{d}x$$

Fick's second law of diffusion expresses the progressive depletion effect with time (as shown in Figure 2.10):

$$\mathrm{d}C/\mathrm{d}t = D\,\mathrm{d}^2C/\mathrm{d}x^2 \tag{2.9}$$

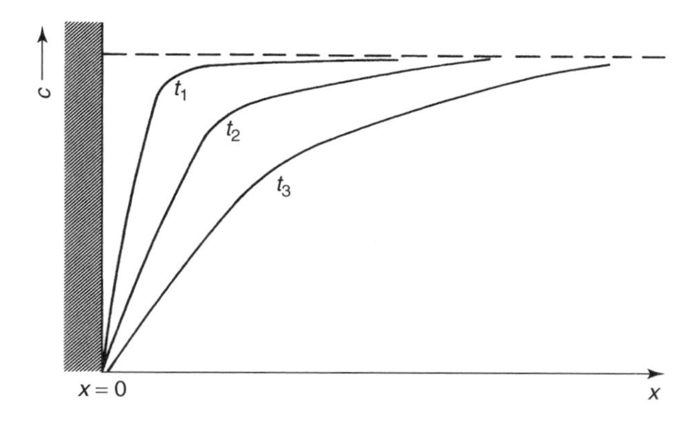

Figure 2.10 The diffusion effect at an electrode for a potential step redox process. From Eggins, B. R., *Biosensors: An Introduction*, Copyright 1996. © John Wiley & Sons Limited. Reproduced with permission.

Obtaining even simple solutions to this equation requires fairly advanced mathematics, and for more complex situations only approximate solutions are possible by using numerical methods of integration (with a computer). However, some of the resulting expressions are fairly straightforward. For our purposes, the most important result is an expression for the current at the peak in Figure 2.9 (point C). For the reversible electron-transfer situation, the peak current (i_p) is given by the Randles–Sevčik equation, as follows:

$$i_p = 2.68 \times 10^5 n^{3/2} A D^{1/2} C_{Ox} v^{1/2} \quad \text{(at 298 K)} \qquad (2.10)$$

If the reverse reaction does not occur, or if the electron-transfer rates are slow (small values of k_f and k_b), the reaction is called *irreversible* and the equation is modified to become as follows:

$$i_p = 2.98 \times 10^5 n (\alpha n_a)^{1/2} A D^{1/2} C_{Ox} v^{1/2} \quad \text{(at 298 K)} \qquad (2.11)$$

The important thing to notice here is that in both cases *the peak current is directly proportional to the concentration of Ox*. This is the most important result for use in analysis.

The potential at which the peak occurs, E_p, is related to the standard redox potential for the couple, Ox/R, by the following equation:

$$E_p = E^0 + 0.056/n$$

for the reversible situation. For the irreversible situation, other factors are also involved.

DQ 2.5

Discuss the factors that contribute to the passage of electric current through an electrolyte solution.

Answer

Three processes can contribute to the flow of electricity through an electrolyte solution. **Migration** *involves the movement of ions across the potential gradient between two electrodes at different potentials. This is studied during conductivity and impedance measurements, but is eliminated during voltammetric studies by the addition of a large concentration of a supporting electrolyte.* **Convection** *involves the movement of the whole solution, carrying the charged particles with it. Stirring or the use of a rotating electrode produces convection. For pure voltammetry, this is avoided by maintaining the cell under quiet, stable conditions.* **Diffusion** *is the movement of material from a high-concentration region of the solution to a low-concentration region. Thus, if the potential at an electrode reduces (or oxidizes) the analyte, its concentration at the electrode surface will be diminished, and so more analyte moves to the electrode from the bulk of the solution. This is the main current-limiting factor in voltammetric processes.*

SAQ 2.8

(a) Explain the term 'diffusion current'.
(b) How is the diffusion current related to the concentration of the reacting analyte?

2.3.2 Cyclic Voltammetry

While the amount of Ox at the electrode surface becomes depleted by the reduction process, it is, of course, replaced by the reduced species (R), which diffuses away into the solution. Hence, if we reverse the potential sweep from the positive side of the peak, we shall observe the reverse effect. As the potential sweeps back towards the redox potential, the reduced species will start to be re-oxidized to Ox. The current will now increase in the negative (oxidizing) direction until an oxidation peak is reached. Figure 2.11 shows the resulting scan of potential against time, while the overall cyclic voltammogram is shown in Figure 2.12.

The latter figure shows two peaks, one corresponding to the reduction of the original substrate and the second corresponding to the re-oxidation of the product back to the original substrate. The peak currents are of almost identical heights. The peak potentials are shifted by $0.056/n$ V relative to each other. The average of the two peak potentials is equal to the standard redox potential, regardless of the concentration of substrate or its diffusion coefficients or rates of electron transfer.

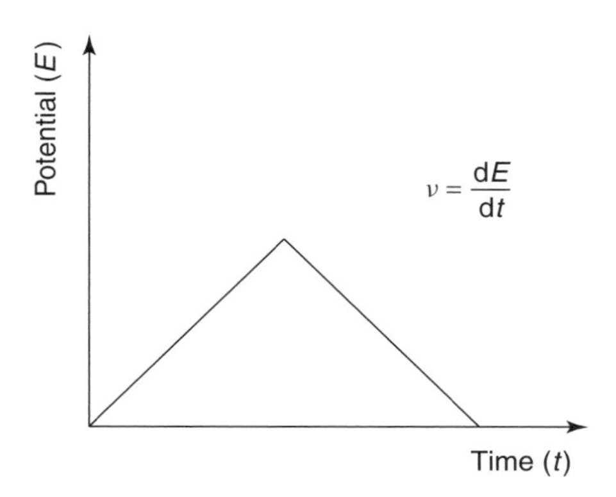

Figure 2.11 A typical scan of potential as a function of time. From Eggins, B. R., *Biosensors: An Introduction*, Copyright 1996. © John Wiley & Sons Limited. Reproduced with permission.

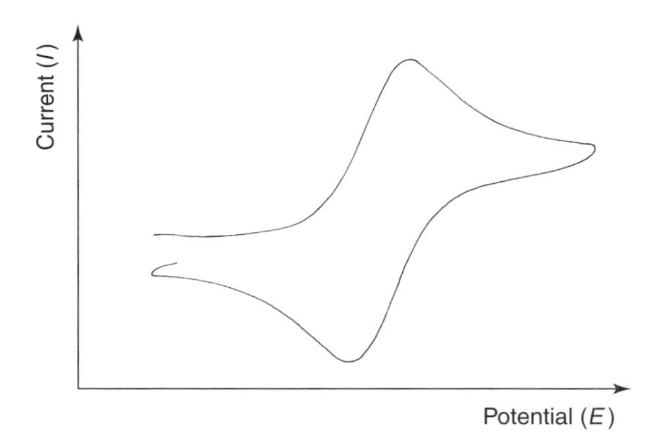

Figure 2.12 A typical reversible cyclic voltammogram. From Eggins, B. R., *Biosensors: An Introduction*, Copyright 1996. © John Wiley & Sons Limited. Reproduced with permission.

If the electrode process is irreversible, the peaks will be shifted away from each other, so that $E_{p(c)} - E_{p(a)} > 0.056/n$. However, the average of the two peak potentials is still a good measure of E^0. Sometimes, the reverse peak will not be present at all or will have a different or distorted shape. This indicates complete irreversibility due to further reaction of the initially formed reduction product. This is best shown by the example of 4-acetylaminophenol (paracetamol), which is easily oxidized electrochemically at a carbon paste electrode. The resulting cyclic voltammogram is shown in Figure 2.13.

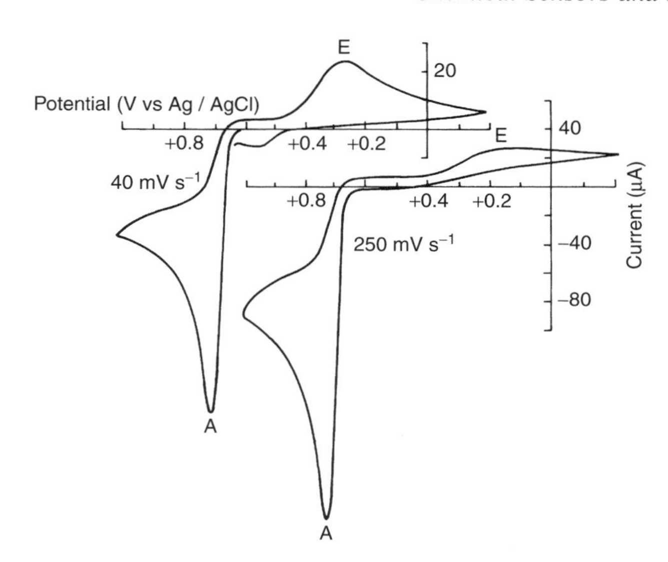

Figure 2.13 The cyclic voltammogram of paracetamol. From Benschoten, J. J., Lewis, J. Y., Heineman, W. R., Roston, D. A. and Kissinger, P. T., *J. Chem Educ.*, **60**, 772–776 (1983). Copyright (1983) American Chemical Society.

SAQ 2.9

(a) In cyclic voltammetry, why is the anodic peak potential not exactly equal to the cathodic peak potential?
(b) How can one obtain a value for the standard redox potential from the two peak potentials?

As the initial reaction is an oxidation process (removal of electrons), the initial peak is downwards. On the reverse scan, the corresponding reduction peak is well separated from the initial oxidation peak. It is also much broader. A study of this system at three different sweep rates and three different pH values elucidates the reaction mechanism and explains the behaviour of the cyclic voltammograms. Figure 2.14 shows the mechanism of this reaction.

2.3.3 Chronoamperometry

A different, but closely related, technique, which clearly shows diffusion control, is *chronoamperometry*. In a modified form it is particularly useful for biosensors. Instead of sweeping the potential, the latter is stepped in a square-wave fashion (Figure 2.15) to a potential just past where the peak would appear in linear-sweep voltammetry. The current is then monitored as a function of time. Decay occurs because of the collapse (or spreading out) of the diffusion layer, as shown above

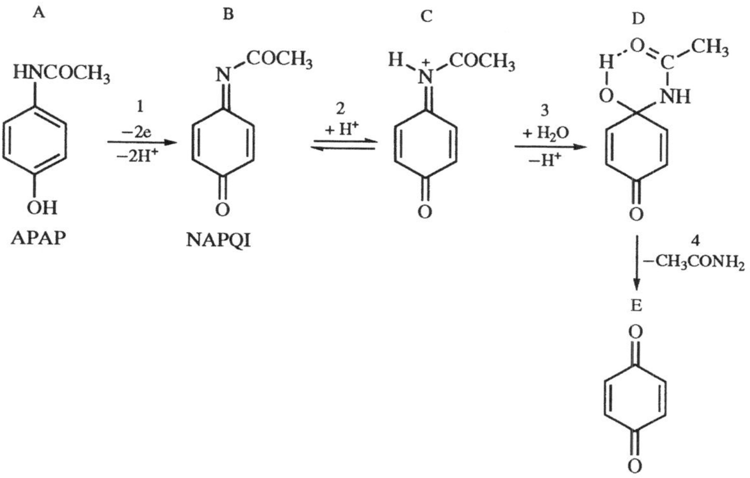

Figure 2.14 The mechanism of the voltammetric oxidation of paracetamol. From Benschoten, J. J., Lewis, J. Y., Heineman, W. R., Roston, D. A. and Kissinger, P. T., *J. Chem Educ.*, **60**, 772–776 (1983). Copyright (1983) American Chemical Society.

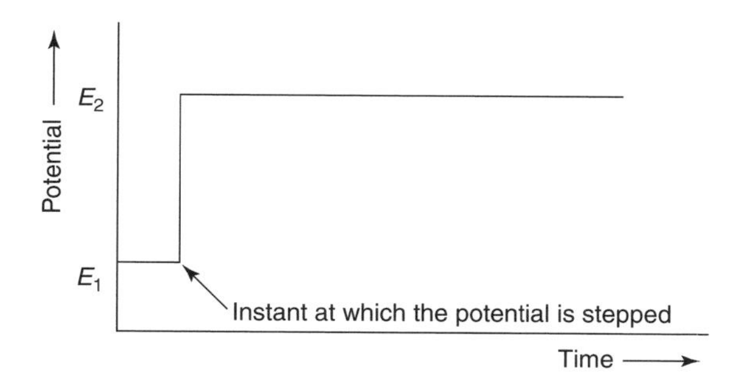

Figure 2.15 Potential step function for chronoamperometry.

in Figure 2.10. The simplest solution to the diffusion equation, which can be obtained analytically, shows that the decay is proportional to the reciprocal of the square root of time, as shown by the Cottrell equation:

$$i_d = \frac{nFADC_{Ox}}{\pi^{1/2}t^{1/2}} \tag{2.12}$$

The current–time profile is shown in Figure 2.16.

Chronoamperometry can be used to determine any of the variables in the above equation, providing that the others are known. Usually, the equation is used to

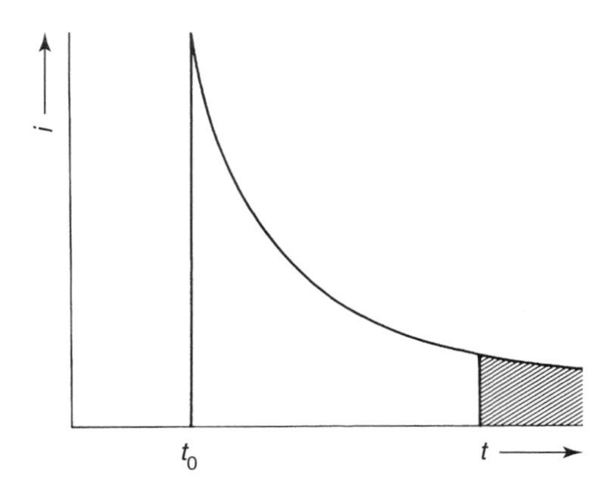

Figure 2.16 The current–time profile associated with a potential step redox process. From Eggins, B. R., *Biosensors: An Introduction*, Copyright 1996. © John Wiley & Sons Limited. Reproduced with permission.

determine n, A or D, plus also sometimes C. The term $i_d/t^{1/2}$ can be determined from the data obtained from Figure 2.16. The different variables can then be determined.

2.3.4 Amperometry

This is the usual name for the analytical application of the chronoamperometric technique. With certain cell and electrode configurations, the decaying current reaches an approximately steady state after a certain time. This is shown by the shaded part of the curve in Figure 2.16. The current has become effectively independent of time, as indicated by the following equation:

$$i = \frac{nFAD^{1/2}C_{Ox}}{\delta}$$

where δ is a constant related to the diffusion layer thickness. This relationship is much more useful for analytical work, even though the current has decayed considerably from its highest values.

SAQ 2.10

How is the diffusion current related to the following:

(a) The concentration of the analyte?
(b) The area of the electrode surface?
(c) The time from the initiation of the potential sweep or step?
(d) The sweep rate in the case of voltammetry?

2.3.5 Kinetic and Catalytic Effects

In most applications involving biosensors, simple reversible or irreversible reactions are rare. Usually, there is a coupled chemical reaction, with maybe just a proton transfer, although sometimes a more complex reaction takes place, as in the example of paracetamol given above in Figures 2.13 and 2.14. The following generalized equations apply:

$$\mathrm{Ox} + n\mathrm{e}^- = \mathrm{R}$$

$$\mathrm{R} + \mathrm{A} \xrightarrow{k} \mathrm{B}$$

Another effect which is particularly useful and common with biosensors is the *catalytic reaction*, in which the original reactant (Ox) is regenerated in the follow-up reaction. This is described by the following equations:

$$\mathrm{Ox} + n\mathrm{e}^- = \mathrm{R}$$

$$\mathrm{R} + \mathrm{A} \xrightarrow{k} \mathrm{Ox} + \mathrm{B}$$

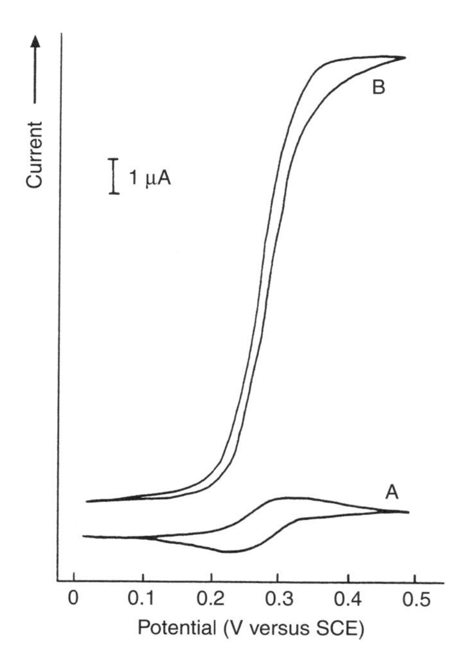

Figure 2.17 Catalytic cyclic voltammograms of ferrocene (as the monocarboxylic acid), obtained (A) in the presence of glucose, and (B) with the addition of glucose oxidase. Reprinted with permission from Cass, A. E. E., Davies, G., Francis, G. D., Hill, H. A, O., Aston, W. J., Higgins, I. J., Plotkin, E. V., Scott, L. D. and Turner, A. P. F., *Anal. Chem.*, **56**, 6567–6571 (1984). Copyright (1984) American Chemical Society.

A detailed analysis of this situation is complicated, although this is not needed for understanding the operation of a biosensor. The effect of this reaction is that the redox process cycles round many times. The reverse oxidation of R is not seen but the forward reduction peak is enhanced many times (see Figure 2.17).

In this figure, the reversible cyclic voltammogram is of ferrocene (dicyclopentadieneiron(III)). The catalytic wave is caused by interaction with glucose oxidase in the presence of glucose (described in more detail below in Chapter 5).

SAQ 2.11

What is the essential difference between a catalytically limited wave and a kinetically limited wave?

2.4 Conductivity

Conductivity is the *inverse of resistance*. It is a measure of the ease of passage of electric current through a solution. Ohm's law gives the following relationship:

$$E = IR$$

and for the conductance, L [in siemens (S), where 1 S = 1 ohm^{-1}]:

$$L = 1/R$$

and therefore:

$$E = I/L$$

Conductance is related to the dimensions of a cell in a similar way to resistance. For a cell of length l and cross-sectional area A, the conductance $L = \kappa A/l$, where κ is the specific conductivity (S cm^{-1}). This is often further normalized by dividing by the molality of the solute to give the molar conductivity, $\Lambda = \kappa/C$ (C in mol cm^{-3}), so the units of Λ are S $\text{mol}^{-1}\text{cm}^{-2}$.

Conductivity is fairly simple to measure, being directly proportional to the concentration of ions in the solution. Figure 2.18 shows a general conductivity bridge circuit. In the traditional bridge, R_3 is adjusted to balance the bridge and a cell constant is then used to convert conductance into (specific) conductivity. In modern instruments, this is carried out automatically to give a digital read-out.

The conductivity varies according to the charge on the ion, the mobility of the ion and the degree of dissociation of the ion. These all introduce complications. In addition, in itself the technique has no selectivity. It can be used in controlled situations but really needs to have selectivity superimposed by means of a membrane or coating.

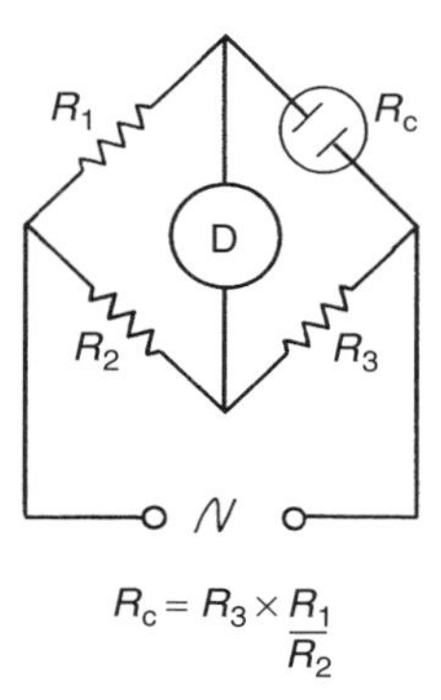

$$R_c = R_3 \times \frac{R_1}{R_2}$$

$$L_c = \frac{1}{R_c} = \frac{R_2}{R_1 R_3}$$

Figure 2.18 A conductivity bridge. From Eggins, B. R., *Biosensors: An Introduction*, Copyright 1996. © John Wiley & Sons Limited. Reproduced with permission.

The measurement of conductance involves an alternating current, as in the classical conductance bridge. Varying the frequency of the alternating current can extend this. The quantity measured is then the *admittance* = 1/*impedance*), which not only depends on simple conductance but also on the capacitance and inductance of the system. These components can be separated as imaginary components, in particular by using a frequency response analyser, and then displaying in an *Argand diagram*, as shown in Figure 2.19. Such diagrams are sometimes called *admittance (impedance) spectra*. This approach has not so far been used to any great extent in developing sensors and biosensors, but is now receiving increased attention.

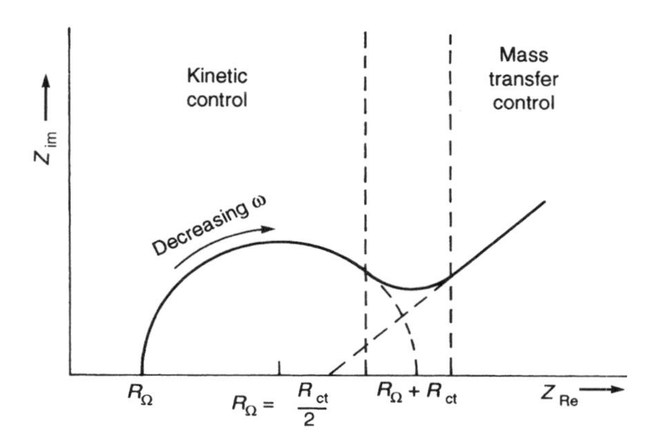

Figure 2.19 A typical Argand diagram, showing the frequency dependence of the 'imaginary' impedance against the 'real' impedance. From Bard, A. J. and Faulkner, L. R., *Electrochemical Methods: Fundamentals and Applications*, © Wiley, 1980. Reprinted by permission of John Wiley & Sons, Inc.

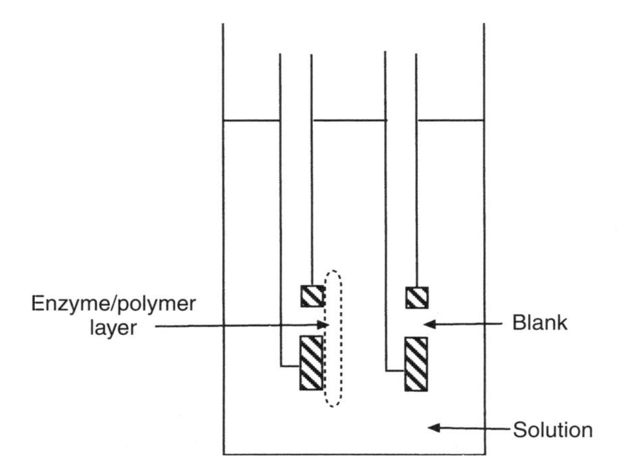

Figure 2.20 Schematic of a differential type of conductivity cell, as used in biosensors. From Eggins, B. R., *Biosensors: An Introduction*, Copyright 1996. © John Wiley & Sons Limited. Reproduced with permission.

SAQ 2.12

Why can direct current not be used in a conductivity bridge?

In principle, a change in conductance can be used to follow any reaction that produces a change in the number of ions, the charge on the ions, the dissociation of the ions or the mobility of the ions. Usually a differential type of cell is used, as shown in Figure 2.20.

DQ 2.6

Discuss factors, which would enable one to use conductance devices as transducers.

Answer

Any reaction or change that involves a change in the number of ions, the charge on the ions or the mobilities of the different ions will produce a change in the conductivity of the solution, which could therefore be used as the transducer. This is a relatively simple, although a somewhat under-used method.

2.5 Field-Effect Transistors

Field-effect transistors (FETs) are devices in which a transistor amplifier is adapted to be a miniature transducer for the detection and measurement of potentiometric signals, produced by a potentiometric sensor process on the gate of the FET. A separate reference electrode is also needed. Circuit wiring is minimized, so that in addition to miniaturization, electronic noise is greatly reduced and sensitivity is increased. The FET device can be part of an integrated-circuit system leading to the read-out, or to the processing of the analytical data. However, as yet, no particularly satisfactory miniaturized reference electrodes exist. According to Janata (see Bibliography), most of the proposed versions violate some of the basic principles of reference electrodes. Despite this, a number of possibilities have been proposed and used, varying from a 'pseudo-reference electrode', consisting of a single platinum or silver wire, to the screen-printed type made with silver–silver chloride ink. Perhaps a more satisfactory approach is to avoid the problem by operating in a differential mode with two FETs, i.e. one being a blank with a gate having negligible response to the analyte and the other coated with the analyte-selective membrane.

2.5.1 Semiconductors – Introduction

Materials can be classified as metals, non-metals or semiconductors. Generally metals are good conductors of electricity, while non-metals are bad conductors i.e. they behave as insulators. Semiconductors come somewhere in between. The differences can be seen in the way that they form energy levels. Non-metal atoms form discrete bonding and anti-bonding molecular orbitals when they combine to form molecules. The bonding orbitals contain the bonding electrons, while the anti-bonding orbitals are empty, unless electrons are promoted into these by excitation. The energy space between these levels is 'forbidden' and is therefore unoccupied. In metals, there are overlapping energy bands and so there is no forbidden region. Electrons can move freely throughout the bands, thus leading to their high conductivities. Semiconductors form energy bands, although in this case they are separated by a forbidden region. The lower band is known as the *valence band* (VB), while the upper band is called the *conduction band* (CB). The energy gap between the two is called the *band gap*. These features are shown in Figure 2.21.

If small amounts of dopants are added to a semiconductor, it may acquire an excess of electrons to give a p-type semiconductor, or a deficit of electrons (excess of holes) giving an n-type semiconductor. Fifth-row elements (in the Periodic Table), such as arsenic, form p-type semiconductors, while third-row elements, like gallium, will form n-type semiconductors.

The Fermi level (E_F) is the point where the probability of filling the (energy) band is 0.5. For an undoped (intrinsic) semiconductor, this will be half-way between the VB and the CB, while for a doped semiconductor, E_F lies nearer to the VB in p-doped materials and nearer to the CB in n-doped materials.

A common arrangement of semiconductors for sensor applications is the metal–insulator–semiconductor (MIS) system. If no potential is applied, the Fermi level is the same across the metal to the semiconductor. However, when a

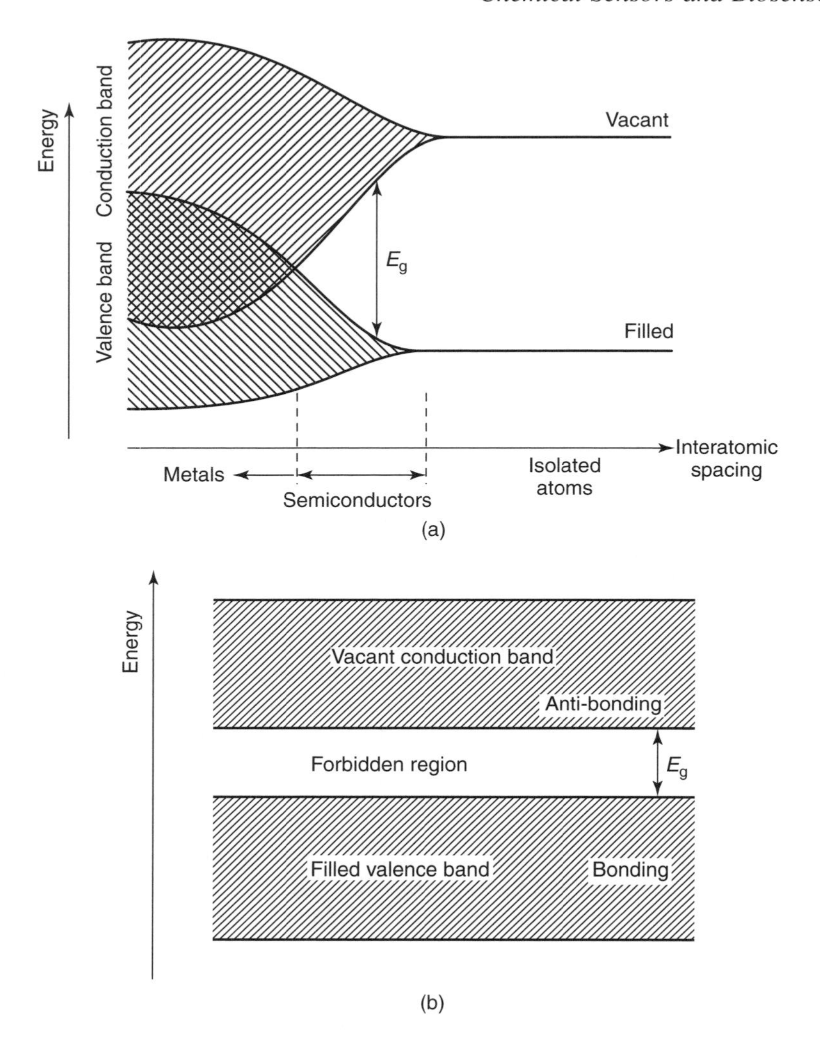

Figure 2.21 (a) Classification of a material according to energy bands and interatomic spacing. (b) The semiconductor band gap energy model. From Hall, E. A. H., *Biosensors*, Copyright 1990. © John Wiley & Sons Limited. Reproduced with permission.

potential is applied the levels on the two sides separate. The system then behaves like a capacitor and charges build up on each side. Figure 2.22 shows the energy levels across a p-type semiconductor. It also shows the effect of applying a potential (a gate voltage, V_G) across the MIS system. With a small negative potential ($V_G < 0$), there is an accumulation of electrons at the metal/insulator (M/I) interface, and of holes (positive charges) at the semiconductor/insulator

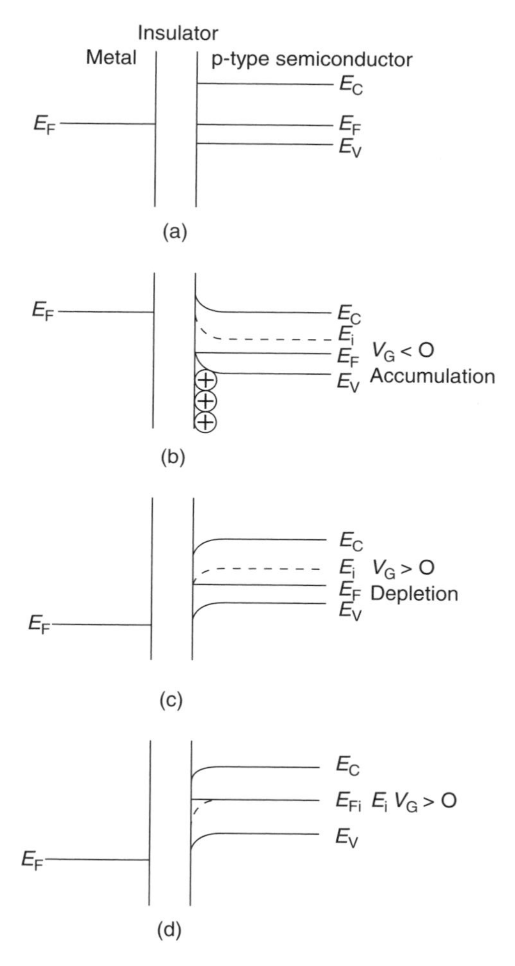

Figure 2.22 Energy bands through an MIS system as a function of the applied voltage V_G. From Hall, E. A. H., *Biosensors*, Copyright 1990. © John Wiley & Sons Limited. Reproduced with permission.

(S/I) interface. E_F is shifted towards the VB lower than the value in the metal by an amount equal to V_G and the energy levels near the semiconductor become bent upwards to compensate for this. With a small positive potential ($V_G > 0$), there is a depletion effect as positive holes are repelled from the S/I interface. In this case, the VB and CB bend downwards to compensate for this. If V_G is further increased, eventually the hole and electron concentrations near the interface become equal. Now, the Fermi level is again midway between the VB and the CB – equivalent to the intrinsic level (Figure 2.22(d)). Further increases in potential beyond this lead to an excess electron concentration, thus causing the semiconductor to invert and become n-type in nature. The potential required to cause inversion is known as the threshold potential (V_T).

2.5.2 Semiconductor–Solution Contact

When an n-type semiconductor is in contact with a solution containing a redox couple (Ox/R), the Fermi level is related to the redox potential E^0. If the E_F of the semiconductor lies above that of the solution, there will be a net flow of electrons from the former into the solution and the CB and VB will be bent upwards (as shown in Figure 2.23). If the interface region is illuminated with light

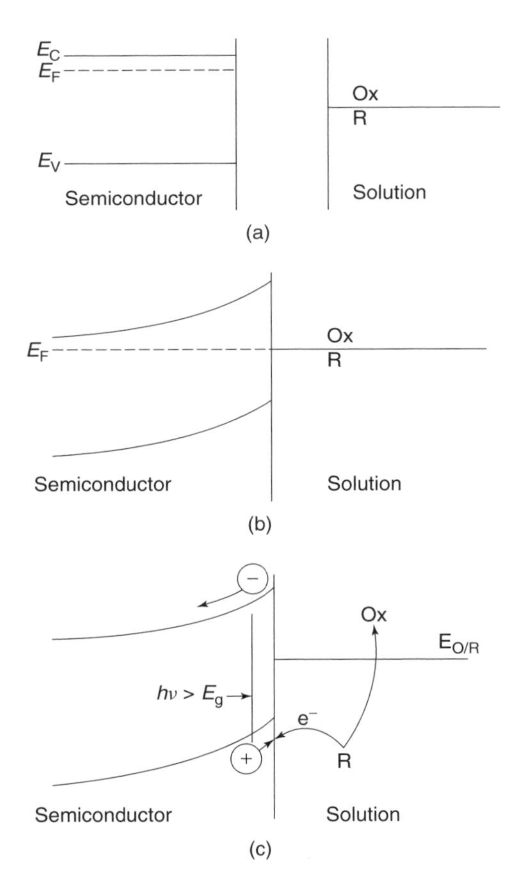

Figure 2.23 Formation of a junction between an n-type semiconductor and a solution containing a redox couple Ox/R: (a) before contact; (b) at equilibrium in the dark; (c) after irradiation, where $h\nu > E_G$. From Hall, E. A. H., *Biosensors*, Copyright 1990. © John Wiley & Sons Limited. Reproduced with permission.

of energy greater than the band-gap energy (E_G), there will be a separation of the electron–hole pairs. The holes migrate to the surface with a potential equivalent to the VB and cause oxidation of R to Ox. The electrons move into the bulk semiconductor and to the external circuit or react with an electron-acceptor (Ox) species, thus casing reduction. This phenomenon is known as photocatalysis. Titanium dioxide is used extensively as a photocatalyst material.

SAQ 2.13

Explain how inversion occurs in a field-effect transistor.

2.5.3 *Field-Effect Transistor*

This is an arrangement to monitor and control changes in the MIS system. Inversion at a p-type S/I system can be monitored by two n-type sensors placed on either side of the p-type layer. The basic type of field-effect transistor (FET) is the insulated-gate FET (IGFET). This is shown in Figure 2.24. A source region (4), consisting of n-type silicon, is separated from a similar drain region (5), also of n-type silicon, by p-type silicon (1), with the insulator (2) consisting of silicon dioxide. The source is electrically biased with respect to the drain by the applied potential, V_D. The gate (3) is a metal. insulated from the rest, so that it

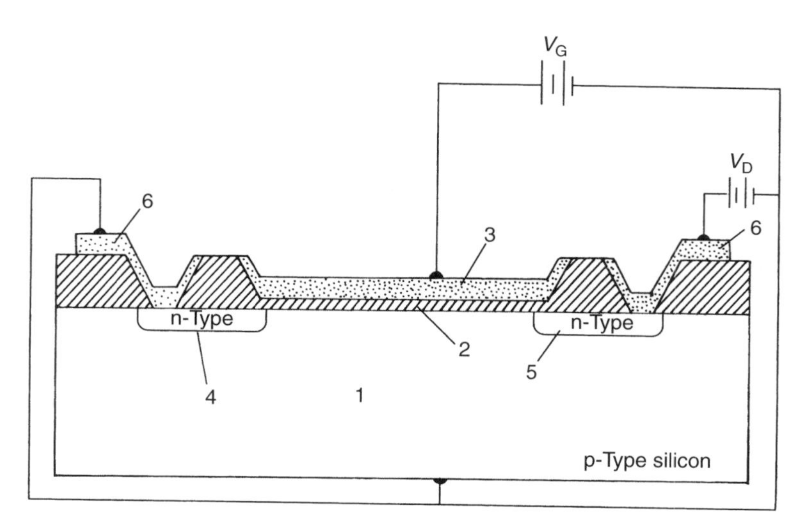

Figure 2.24 Schematic of the insulated-gate field-effect transistor (IGFET): 1, p-type silicon substrate; 2, insulator; 3, gate metal; 4, n-type source; 5, n-type drain; 6, metal contacts to source and drain. From Eggins, B. R., *Biosensors: An Introduction*, Copyright 1996. © John Wiley & Sons Limited. Reproduced with permission.

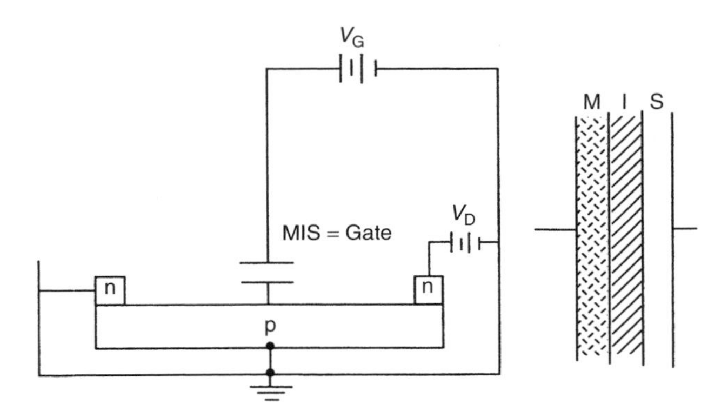

Figure 2.25 Schematic of the gate in an IGFET: M, metal; I, insulator; S, semiconductor. From Eggins, B. R., *Biosensors: An Introduction*, Copyright 1996. © John Wiley & Sons Limited. Reproduced with permission.

forms a capacitor sandwich, a metal/insulator/semiconductor (MIS) arrangement, as shown in Figure 2.25

This gate region is charged with a bias potential V_G,. The current from the drain (5) to the source (4), I_D, is measured. There is also a threshold potential, V_T, at which silicon changes from p-type to n-type, and inversion occurs. With a small positive V_D and $V_G < V_T$, silicon (1) remains in the p-state, and there is no drain current; n-Si is biased positive with respect to p-Si. When $V_G > V_T$, there is surface inversion, and p-Si becomes n-Si. Now current can pass from drain to source, without crossing the reversed-bias p–n junction. V_G now modulates the number of electrons from the inversion layer and so controls the conductance. I_D flows from source to drain, and is proportional to both the electrical resistance of the surface inversion layer and V_D.

In order to convert this device into a sensor, the metal of the gate is replaced by a chemically sensing surface. This general conformation is known as a CHEMFET and is shown in Figure 2.26.

In this arrangement, the chemically sensitive membrane (3) is in contact with the analyte solution (7). A reference electrode (8) completes the circuit via the V_G bias. The membrane potential minus the solution potential has the effect of correcting for this bias.

The current may be measured directly at constant V_G by using a circuit such as that shown in Figure 2.27. Alternatively, one can keep I_D constant by changing V_G and measuring the latter by using a circuit such as the arrangement shown in Figure 2.28. Such a system is used in a number of sensor modes. The general CHEMFET mode has already been mentioned. A further mode is the ion-selective mode (ISFET), which uses the FET as an ion-selective electrode. Following on from this, the ENFET is a form of biosensor in which the gate contains an enzyme system.

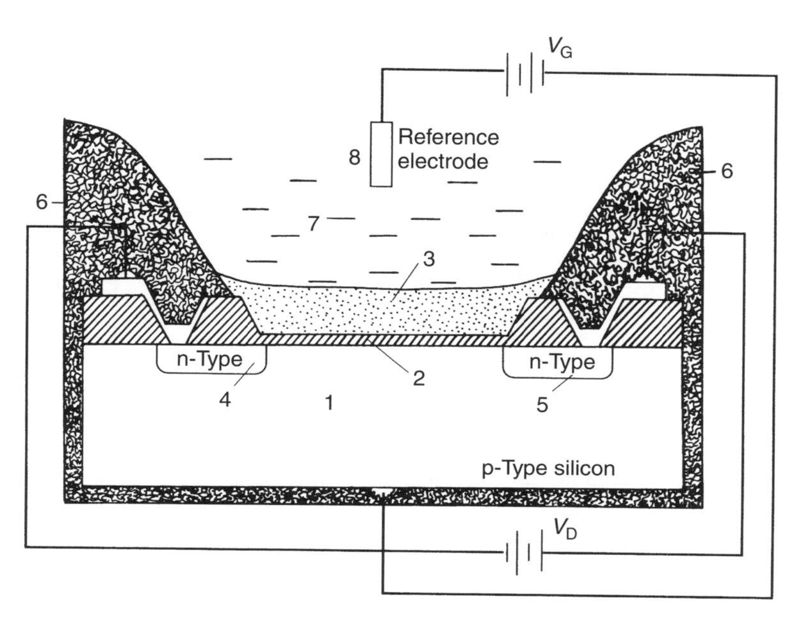

Figure 2.26 Schematic of a field-effect transistor with a chemically sensing gate surface (CHEMFET): 1, silicon substrate; 2, insulator; 3, chemically sensitive membrane; 4, source; 5, drain; 6, insulating encapsulant; 7, analyte solution; 8, reference electrode. From Eggins, B. R., *Biosensors: An Introduction*, Copyright 1996. © John Wiley & Sons Limited. Reproduced with permission.

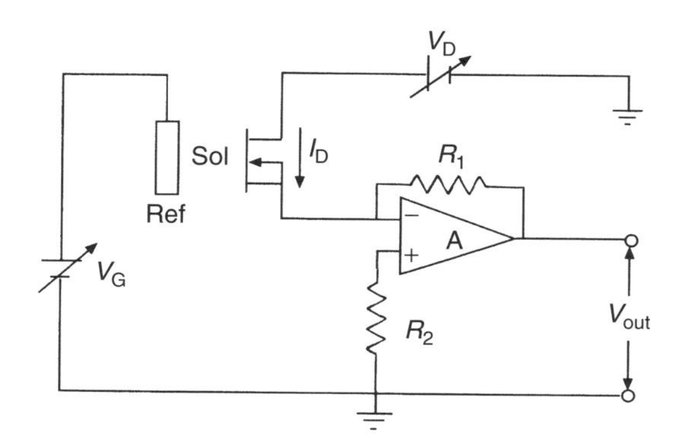

Figure 2.27 Schematic of the circuit used for measuring I_G at a constant gate voltage: A, operational amplifier; R_1, 1 kΩ; R_2, 470 Ω. From Eggins, B. R., *Biosensors: An Introduction*, Copyright 1996. © John Wiley & Sons Limited. Reproduced with permission.

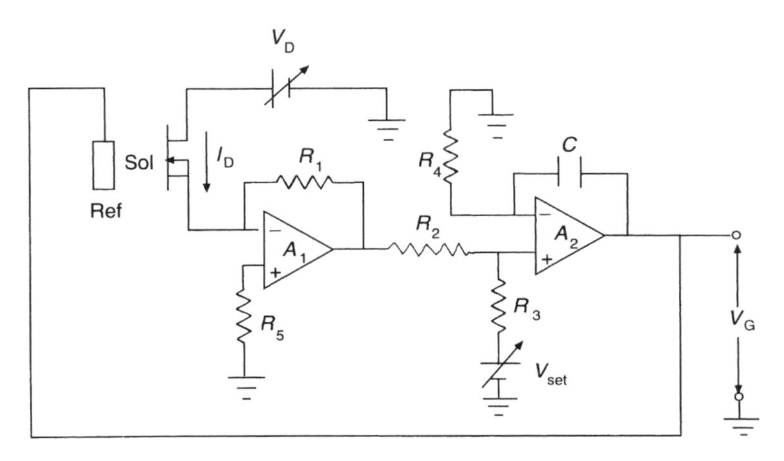

Figure 2.28 Schematic of the circuit used for measuring changes in V_G at a constant drain current: A_1 and A_2, operational amplifiers; R_1, 1 KΩ; $R_2 = R_3$, 100 kΩ; R_4, 20 KΩ; R_5, 470 Ω; *c*, 10 pF. From Eggins, B. R., *Biosensors: An Introduction*, Copyright 1996. © John Wiley & Sons Limited. Reproduced with permission.

2.6 Modified Electrodes, Thin-Film Electrodes and Screen-Printed Electrodes

Modified electrodes will be discussed in detail in Chapter 3.

A major aspect in the manufacture of sensors is miniaturization. Three developments, which have assisted this, are thick-film electrodes formed by screen-printing, thin-film electrodes and microelectrodes.

2.6.1 Thick-Film – Screen-Printed Electrodes

Here, the working electrode is usually a graphite-powder-based 'ink' printed on to a polyester material. The reference electrode is usually silver–silver chloride ink. A typical layout is shown in Figure 2.29.

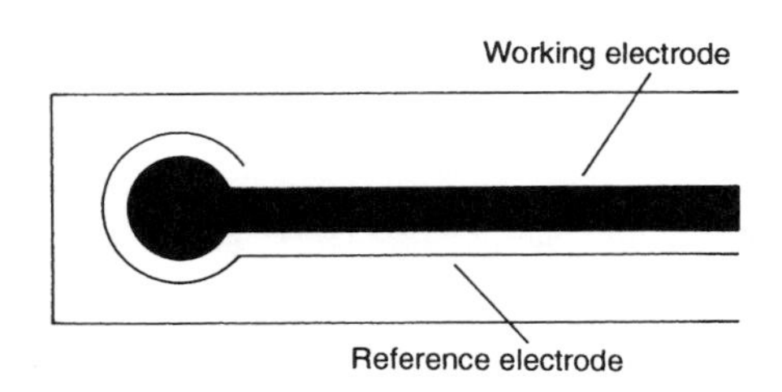

Figure 2.29 A screen-printed electrode. From Wang, J., *Analyst*, **119**, 763–766 (1994). Reproduced with permission of The Royal Society of Chemistry.

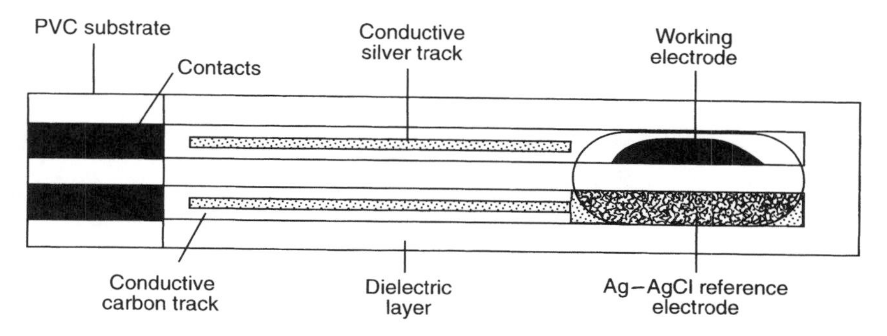

Figure 2.30 Schematic of the ExacTech biosensor disposable electrode strip. From Hilditch, P. I. and Green, M. J., *Analyst*, **116**, 1217–1220 (1991). Reproduced with permission of The Royal Society of Chemistry.

Appropriate modifying components can be incorporated into the carbon ink, such as gold, mercury, chelating agents (for use in stripping voltammetry), mediators such as phthalocyanines and ferrocenes to catalyse electron transfer, or enzymes such as glucose oxidase, ascorbic acid oxidase, glutathione oxidase or uricase. The procedure has the advantages of miniaturization, versatility and cheapness, and in particular lends itself to the mass production of disposable electrodes. A version is marketed commercially in the 'ExacTech' biosensor for glucose (Figure 2.30).

SAQ 2.14

How could a screen-printed electrode be made by using a plant tissue material such as that of a banana?

2.6.2 Microelectrodes

Microelectrodes, also called ultra-microelectrodes, having dimensions in the range 1–10 μm, have greatly extended the range of sample environments and experimental time-scales that can be used for electroanalysis. Such electrodes have surface areas which are many times smaller than the cross-sectional area of a human hair. They operate with small currents in the pA to nA range, and have steady-state responses and short response times. Electrodes have been made in the form of discs, bands, cylinders, rings and arrays. A simple disc can be made by embedding a platinum wire or carbon fibre in glass or epoxy resin and exposing the (cross-sectional) disc to the solution.

Because of their small dimensions, the double-layer capacitance is low, so that the faradaic current is large when compared to the background capacitive current. Due to the small current magnitudes, the IR drop is very much reduced (or

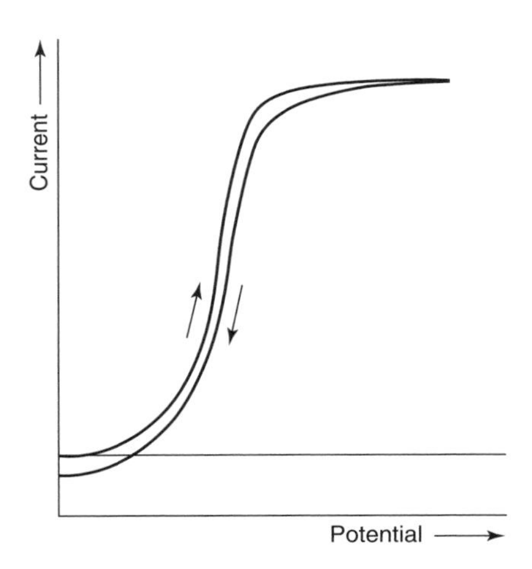

Figure 2.31 An example of a cyclic voltammogram obtained at a microelectrode. From Diamond, D., (Ed.), *Principles of Chemical and Biological Sensors*, © Wiley, 1998. Reproduced by permission of John Wiley & Sons, Inc.

sometimes eliminated), thus allowing easier operation in media of high resistance, such as organic solvents of low polarity.

The voltammetric behaviour of these electrodes is dominated by the 'edge effect'. The result is that a typical cyclic voltammogram has a sinusoidal shape, as shown in Figure 2.31.

A steady-state current results, as defined by the following equation:

$$I_{ss} = 4nFC^{b}Dr$$

where r is the radius of the disc and D the diffusion coefficient of the electroactive species. Note that there is no time-dependence.

Because of the minimal IR 'problem', such electrodes can be used in two-electrode configurations with a conventional reference electrode. The main problems are those of polishing and manipulating the electrodes and the need for a low-noise, low-current measuring device.

Applications in sensor devices include *in vivo* heavy-metal analysis, detectors for high performance liquid chromatography (HPLC) and their use in flowing streams for flow-injection analysis (FIA). They are also used in clinical analyses, especially for brain chemistry. Scanning electrochemical microscopy uses microelectrodes, while arrays have been used to generate 'chromatovoltammograms'.

Again, because of the low IR errors, much faster scan rates can be used than in conventional cyclic voltammetry (700 Vs^{-1}, compared with 0.1 Vs^{-1}), thus allowing enhanced sensitivity because of larger currents and shorter analysis times. They can also be used in capillary electrophoresis to study, for example, ultra-small biological environments for the determination of sub-attomole concentrations of seretonin in a single nerve cell of the pond snail, *Planorbis corneus*.

2.6.3 Thin-Film Electrodes

Thin-film techniques are used to make integrated silicon wafers for 'smart' electrochemical sensors. Thin-layer cells using thin-film electrodes allow rapid complete oxidation or reduction of small volumes of solution. This permits rapid coulometric analyses.

Solid-state gas sensors utilize thin-film electrodes, e.g. a zinc oxide thin film will detect carbon dioxide at 400°C. Adsorption of the gas on to its surface results in a change of conductivity, which can then be measured. Such a sensor will also detect various hydrocarbons, such as toluene and propane, as well as ethyl alcohol and diethyl ether, down to ppm levels. Tin oxide behaves in a similar way. The sensor shown in Figure 2.32, consisting of palladium-doped tin oxide (0.3 μm thick), is grown on top of a silica layer on top of a ferrite substrate. On the other face of the ferrite is a thick layer of ruthenium dioxide. Gold contacts are added to complete the system. These sensors are designed to detect carbon monoxide and ethanol.

Both p-type and n-type semiconductors (mainly the latter) change their resistance from high to low levels in the presence of a gas. The gas is effectively acting here as an electron donor.

The semiconductor adsorbs a small amount of oxygen, which reacts with the excess electrons in the semiconductor, according to the following:

$$O_2 + 2e^- \longrightarrow 2O^-_{ads}$$

This surface oxide then reacts with the gas, as follows:

$$G + O^-_{ads} \longrightarrow GO_{des} + e^-$$

This effect is detected as an increase in the electrical conductivity of the tin oxide.

SAQ 2.15

What are the advantages of ultra-microelectrodes?

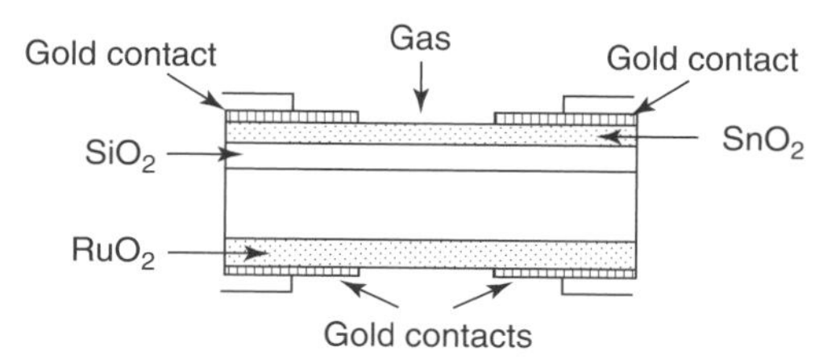

Figure 2.32 Schematic of a tin oxide-based thin-film gas sensor. © R. W. Catterall 1997. Reprinted from **Chemical Sensors** by R. W. Catterall (1997), by permission of Oxford University Press.

2.7 Photometric Sensors

2.7.1 Introduction

Most bioassays were originally of the photometric type. Such assays made use of changes in a species which involved a strong change in the photometric properties. Probably the best known of these is the involvement of NAD^+/NADH in biochemical reactions (where NAD^+ and NADH are, respectively, the oxidized and reduced forms of nicotinamide–adenine dinucleotide). For example:

$$\text{pyruvate} + \text{NADH} + \text{H}^+ = \text{L-lactate} + \text{NAD}^+$$

NADH has a strong absorbance, with a λ_{max} at 340 nm, while NAD^+ has no absorbance at this wavelength. NADH also shows fluorescence at 400 nm. The absorption spectra of NADH and NAD^+ are shown in Figure 2.33.

Such behaviour may be used in the conventional way, in which the reagents and analyte are measured out and mixed and then placed in a cuvette in a

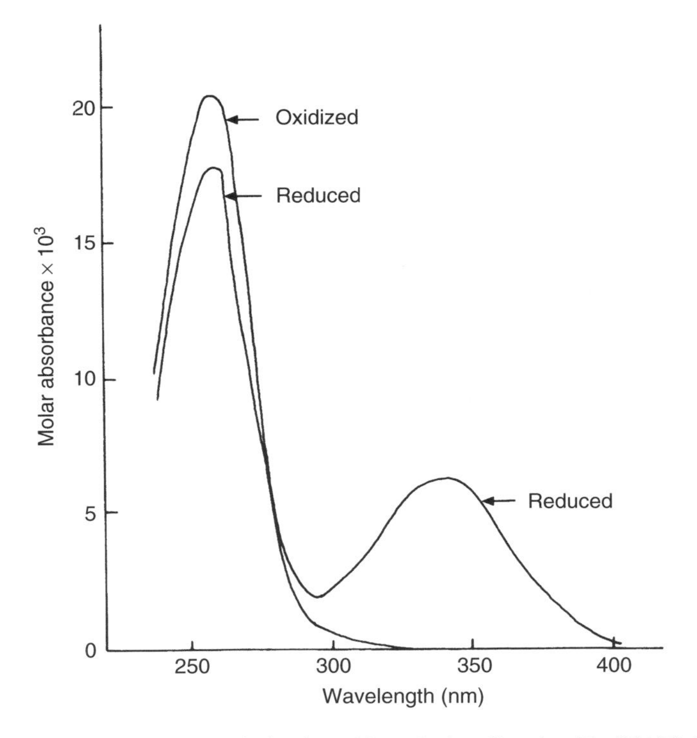

Figure 2.33 Absorption spectra of nicotinamide–adenine dinucleotide (NAD) in its oxidized and reduced forms. From Hall, E. A. H., *Biosensors*, Copyright 1990. © John Wiley & Sons Limited. Reproduced with permission.

spectrophotometer. The problem is how to make a sensor by using an optical technique.

The basic optical response is based on the Beer–Lambert law (usually referred to as Beer's law), as follows:

$$\log(I/I_0) = A = \varepsilon Cl \tag{2.13}$$

where I_0 is the intensity of the incident light, I the intensity of the transmitted light, A the absorbance (usually measured directly by an instrument), ε the extinction coefficient, C the concentration of the analyte, and l the pathlength of light through the solution. The latter can limit the size of sensors, a feature which does not occur in electrochemical devices.

DQ 2.7

What are the advantages of optical sensors?

Answer

The advantages of photometric devices are as follows:

(i) No 'reference electrode' is needed, although a reference source is often useful.

(ii) There is no electrical interference.

(iii) An immobilized reagent does not have to be in contact with any optical fibres, and can easily be replaced.

(iv) There are no electrical safety hazards.

(v) Some analytes, such as oxygen, can be detected in equilibrium.

(vi) They are highly stable with respect to calibration, especially if the ratio of the intensities at two different wavelengths can be measured.

(vii) They can respond simultaneously to more than one analyte by using multiple immobilized reagents with different wavelengths for response, e.g. O_2 *and CO.*

(viii) Multi-wavelength measurements can be made to monitor changes in the state of the reagent.

(ix) They have potential for higher-information content than electrical transducers.

The disadvantages of optical sensors are as follows:

(i) They will only work if appropriate reagent phases can be developed.

(ii) They are subject to background ambient light interference. This may be excluded either directly or by using a modulation technique.

(iii) They have a limited dynamic range when compared to electrical sensors – typically 10^2 compared with 10^6–10^{12} for ion-selective electrodes.

(iv) They are extensive devices, and dependent on the amount of reagent, and hence difficult to miniaturize.

(v) There are problems with the long-term stability of the reagents under incident light.

(vi) Response times may be slow because of the time of mass transfer of analytes to the reagent phase.

2.7.2 Optical Techniques

Let us now survey the range of optical techniques that could potentially be used in such sensors. Figure 2.34 shows the electromagnetic spectrum, illustrating the relative wavelengths of some of the important types of radiation, such as visible, ultraviolet and infrared

The main types of photometric behaviour which have been exploited in biosensor applications are as follows:

- Ultraviolet–visible absorption
- Fluorescence (and phosphorescence) emission
- Bioluminescence
- Chemiluminescence
- Internal reflection spectroscopy (IRS)
- Laser light scattering methods

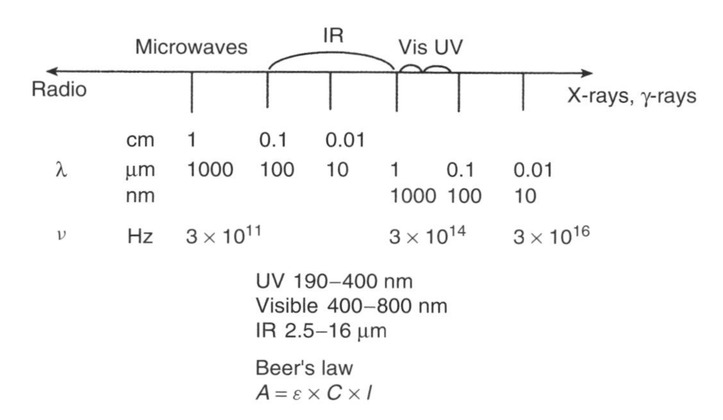

Figure 2.34 The electromagnetic spectrum. From Eggins, B. R., *Biosensors: An Introduction*, Copyright 1996. © John Wiley & Sons Limited. Reproduced with permission.

2.7.3 Ultraviolet and Visible Absorption Spectroscopy

The electronic transitions of organic molecules and biomolecules occur in the ultraviolet or visible regions of the electromagnetic spectrum. The major modes of excitation are as follows:

- $\sigma \rightarrow \sigma^*$ and $\pi \rightarrow \pi^*$, i.e. from bonding to anti-bonding molecular orbitals
- $n \rightarrow \sigma^*$ and $n \rightarrow \pi^*$, i.e. from non-bonding to anti-bonding molecular orbitals

While such spectra can supply some information about the molecular structure, the major application is quantitative analysis, making use of the Beer–Lambert law (see equation (2.13) above).

The absorbance is measured by passing incident radiation through the monochromator of a spectrometer and then measuring its intensity with a photomultiplier or photodiode, thus producing an electrical signal which is proportional to the absorbance at a particular wavelength. The measured absorbance is linearly proportional to the concentration. However, the apparatus required for this is relatively cumbersome and can be expensive, thus limiting the possibilities for exploitation in sensors.

SAQ 2.16

State the Beer–Lambert law.

SAQ 2.17

A 1 cm cell contained a solution of an analyte which was known to absorb at 470 nm, with an extinction coefficient of 10^3. The absorbance was set to zero with water in the cell. When the analyte was placed in the cell in a spectrophotometer, the absorbance was 0.850. What is the concentration of the analyte?

2.7.4 Fluorescence Spectroscopy

After absorption of light, the excited species will decay in one of a variety of ways. If the molecule has a suitably complex structure so that there is time for a radiationless decay to the lowest excited singlet state, it can then re-emit radiation, usually at a lower wavelength than the original excitation. This phenomenon is known as *fluorescence*, and is illustrated in Figure 2.35.

The emitted fluorescent radiation can be measured in a similar way to UV–visible radiation, and is also subject to the Beer–Lambert law. The method used is often more sensitive than absorption spectroscopy. In addition, the change of wavelength has advantages in analysis.

However, while the excited molecule is undergoing radiationless decay, it may also have time to react with another species, thus transferring its energy to the new

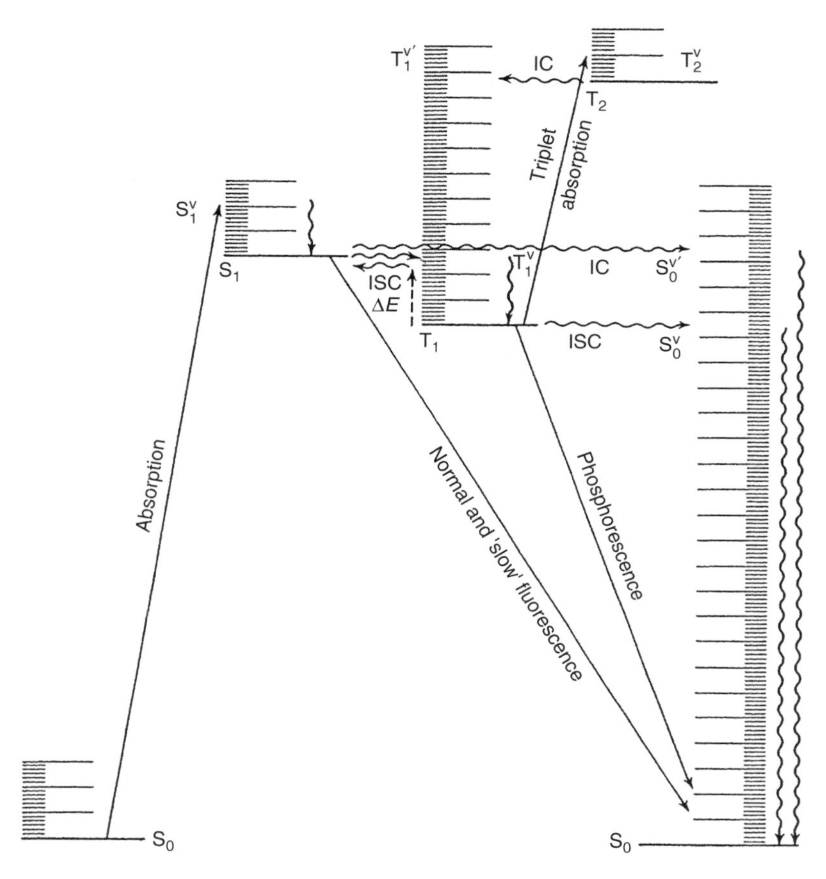

Figure 2.35 Jablonski diagram showing adsorption, and the emission process of fluorescence (the corresponding process of phosphorescence is also indicated). From Wayne, R. P., *Photochemistry*, Butterworths, London, 1970.

chemical species and returning to the ground state with a diminished fluorescent output. This process is known as *quenching* and provides another powerful way of using fluorescent behaviour:

Fluorescence $$M^* \xrightarrow{k_f} M + h\nu$$

Quenching $$M^* + Q \xrightarrow{k_q} M + Q^*$$

The fluorescent intensity is decreased to I_f from the initial absorption intensity I_0. The quenching effect is described by the Stern–Volmer relationship, as follows:

$$I_f/I_0 = 1/(1 + (k_q/k_f)[Q]) \qquad (2.14)$$

DQ 2.8

How does the Beer–Lambert law apply to fluorescence?

Answer

The absorption of light leading to fluorescence obeys the Beer–Lambert law and so does the intensity of the fluorescence. The full expression relating the fluorescent intensity to the concentration of the analyte is as follows:

$$I_f = 2.303\,\phi_f I_0 \varepsilon C l$$

where I_f *is the fluorescence intensity* ϕ_f *the quantum efficiency of the fluorescence process,* I_0 *the intensity of the excitation light,* ε *the molar absorptivity,* C *the concentration of the analyte, and* l *the pathlength of the cell.*

2.7.5 Luminescence

Certain chemical reactions involve the emission of light without heat – this is known as *chemiluminescence*. The similar emission of light from biological systems is called *bioluminescence*.

2.7.5.1 Chemiluminescence

Chemiluminescence occurs as a result of the oxidation of certain substances, usually with oxygen or hydrogen peroxide, producing visible light 'in the cold' and in the absence of any exciting illumination, i.e. in the dark. The best known of these substances is luminol, shown in Figure 2.36. There are other more complex substances which will react in a similar way.

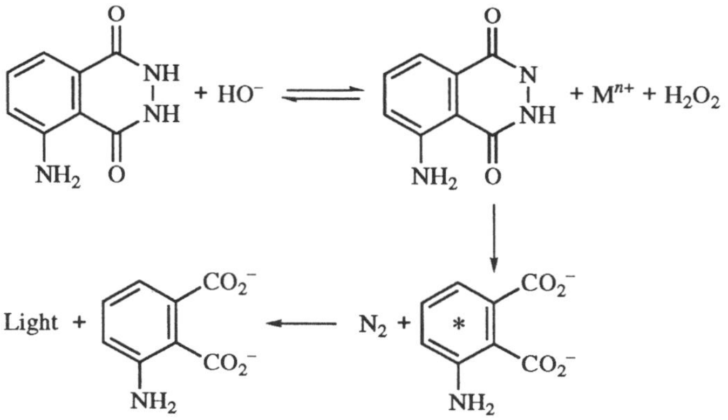

Figure 2.36 Reaction mode of luminol. Reprinted from **Biosensors: Fundamentals and Applications** edited by A. P. F. Turner, I. Karube and G. S. Wilson (1987), by permission of Oxford University Press.

Table 2.1 Some examples of chemiluminescent assay systems. From Hall, E. A. H., *Biosensors*, Copyright 1990. © John Wiley & Sons Limited. Reproduced with permission

Analyte	Label	Chemiluminescent reactants
Human IgG	Luminol	H_2O_2–haemin
Testosterone	Luminol derivative	H_2O_2–Cu(II)
Thyroxine	Luminol derivative	Microperoxidase
Biotin	Isoluminol	H_2O_2–lactoperoxides
Hepatitis B	Isoluminol derivative	Microperoxidase–peroxide
Rabbit IgG	Isoluminol	Microperoxidase–peroxide
Cortisol	Isoluminol	Microperoxidase–peroxide

The luminol is normally used as a *label*. This can be employed in any assay involving oxygen, hydrogen peroxide or peroxidase. It is particularly useful with immunoassays, with Table 2.1 showing some examples of these. However, the sensitivity is limited because the quantum yield is only 1%.

A particularly interesting approach, combining both luminescence and fluorescence, is shown in Figure 2.37. In this, an antigen is labelled with the luminol, while the corresponding antibody is labelled with a fluorescent compound such that the emission from the luminol will excite the fluorescence.

In the bound Ag–Ab, the luminol emits light of 460 nm, which excites the fluorescor, emitting fluorescence in turn at 525 nm, and thus resulting in an increased quantum yield. At the same time, unlabelled antigen may combine with labelled antibody, in which case there is no fluorescence but simply emission from the luminol at 460 nm. This permits analysis of both bound and unbound antigen at the two wavelengths.

Biosensors may be constructed which involve the use of luminol with hydrogen peroxide and peroxidase. A fibre-optic sensor for H_2O_2 can be made from peroxidase immobilized on a polyacrylamide gel containing luminol at the end

Figure 2.37 Competitive immunoassay employing a fluorescent-labelled antibody (Ab) and a chemiluminescent-labelled antigen (Ag). The fluorophore acts as the acceptor of the chemiluminescence energy in the doubly labelled complex. Reprinted from **Biosensors: Fundamentals and Applications** edited by A. P. F. Turner, I. Karube and G. S. Wilson (1987), by permission of Oxford University Press.

of the fibre. The luminescence is detected *in situ* without any diffusion of the luminescent species and, if $[S] \gg K_m$, is independent of the thickness of the membrane. Of course, no external light source is needed. The sensor can be connected directly to the photodiode. It will detect 1–10 mM H_2O_2, with a response time of 2 min. One obvious application is to 'connect' the sensor to a glucose–glucose oxidase reaction system to determine glucose, for which a linear concentration range of 0 15–1.5 mM can be obtained.

A recent example of a new type of chemiluminescence system involves adamantyl dioxetine phosphate (Figure 2.38) which can be hydrolysed under the

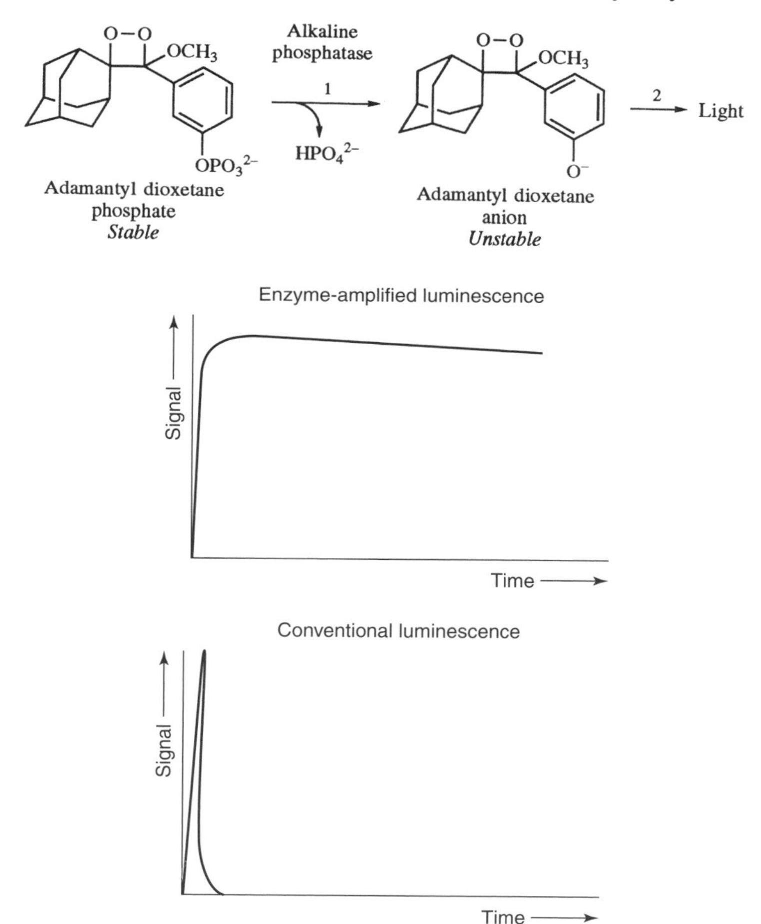

Figure 2.38 Mechanism of alkaline-phosphatase-activated luminescence through the hydrolysis of the phosphate ester of adamantyl dioxetine. From Eggins, B. R., *Biosensors: An Introduction*, Copyright 1996. © John Wiley & Sons Limited. Reproduced with permission.

influence of alkaline phosphatase to form the adamantyl dioxetine anion, which is unstable and fluoresces. The fluorescence lifetime is several minutes, unlike the more conventional type of luminescence. The structures of the materials involved and the mechanism of the fluorescence process are shown in Figure 2.38.

Such behaviour could be used in many types of assay which involve phosphate ester hydrolysis using alkaline phosphatase. The adamantyl dioxetane phosphate could be allowed to compete with other organic phosphates for the phosphatase.

2.7.5.2 Bioluminescence

Certain biological species, principally the firefly, can emit luminescence. This originates in a group of substances of varied structures known as *luciferins* (see Figure 2.39).

The enzyme-catalysed oxidation of luciferin results in luminescence, according to the following:

$$\text{luciferin} \xrightarrow{\text{luciferase, } O_2} \text{oxyluciferin} + h\nu \text{ (562 nm)}$$

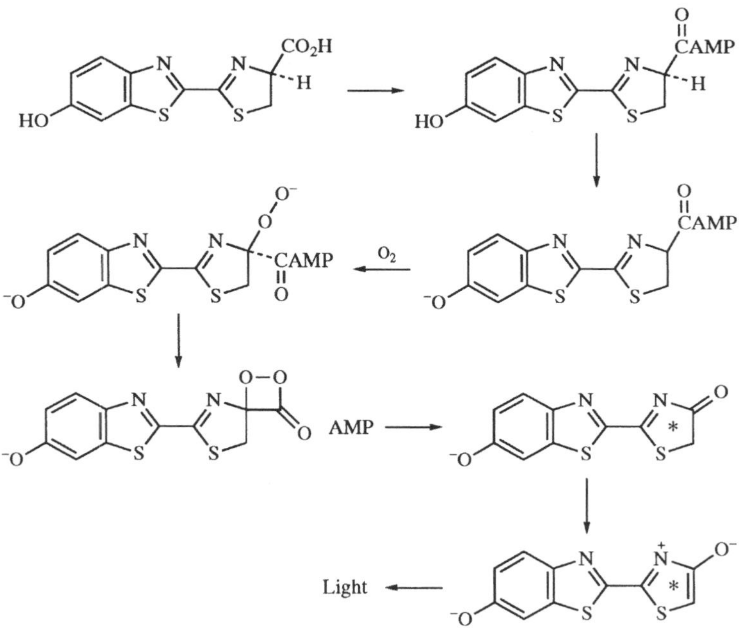

Figure 2.39 Mechanism of the (luciferase) coupling reaction of luciferin with adenosine 5′-triphosphate (ATP). Reprinted from **Biosensors: Fundamentals and Applications** edited by A. P. F. Turner, I. Karube and G. S. Wilson (1987), by permission of Oxford University Press.

Some of the luciferin species will couple with cofactors such as adenosine 5′-triphosphate (ATP), flavin mononucleotide (FMN) and the reduced form of the coenzyme, flavin–adenine dinucleotide (FADH), for example:

$$\text{ATP} + \text{luciferin} + \text{O}_2 \xrightarrow{\text{luciferase}} \text{AMP} + \text{PP} + \text{oxyluciferin} + \text{CO}_2 + \text{H}_2\text{O} + h\nu$$

where PP is pyrophosphate.

This reaction is very sensitive down to femtomole concentrations. The determination of creatine kinase, which is related to the diagnosis of myocardial infarction and muscle disorders, is one important clinical assay that can be carried out by this method:

$$\text{AMP} + \text{creatine phosphate} \xrightarrow{\text{creatine kinase}} \text{ATP} + \text{creatine}$$

where AMP is adenosine 5′-phosphate.

Bacterial luciferases do not involve luciferins but form an excited complex with reduced flavins, such as FMNH, as follows:

$$\text{FMNH}_2 + \text{O}_2 + \text{RCHO} \longrightarrow \text{FMN} + \text{RCOOH} + \text{H}_2\text{O} + h\nu \ (478\text{–}505 \text{ nm})$$

Most analytical reactions involve the reduced form of nicotinamide–adenine dinucleotide (NADH), as shown in Figure 2.40. This could be coupled, for example, to determine ethanol, as follows:

$$\text{ETOH} + \text{NAD}^+ \longrightarrow \text{CH}_3\text{CHO} + \text{NADH} + \text{H}^+$$

as illustrated in Figure 2.41.

A very dramatic application is the determination of trinitrotoluene (TNT) using this system via an immunoassay, as shown in Figure 2.42.

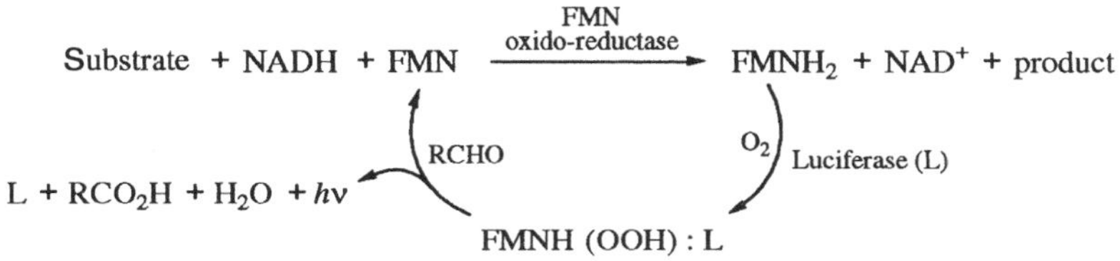

$$\text{Substrate} + \text{NADH} + \text{FMN} \xrightarrow[\text{oxido-reductase}]{\text{FMN}} \text{FMNH}_2 + \text{NAD}^+ + \text{product}$$

O_2, Luciferase (L)

FMNH (OOH) : L

RCHO

$\text{L} + \text{RCO}_2\text{H} + \text{H}_2\text{O} + h\nu$

Figure 2.40 Basic mode of operation of bacterial luciferases via nicotinamide–adenine dinucleotide (NAD) in its reduced form and flavin mononucleotide (FMN). From Hall, E. A. H., *Biosensors*, Copyright 1990. © John Wiley & Sons Limited. Reproduced with permission.

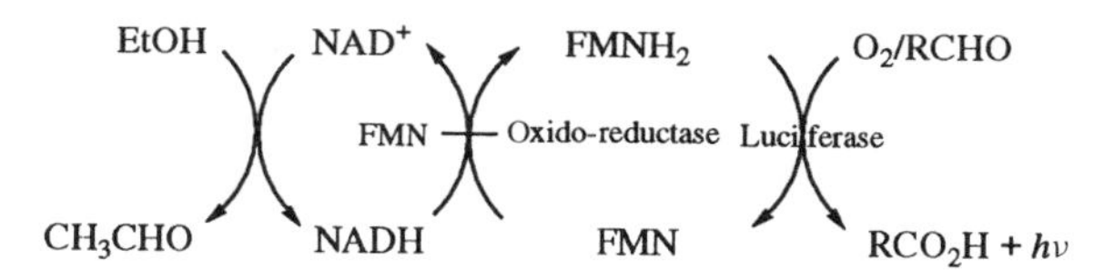

Figure 2.41 Bioluminescent assay of ethanol via NAD and NADH (EtOH/NAD/luciferins). From Hall, E. A. H., *Biosensors*, Copyright 1990. © John Wiley & Sons Limited. Reproduced with permission.

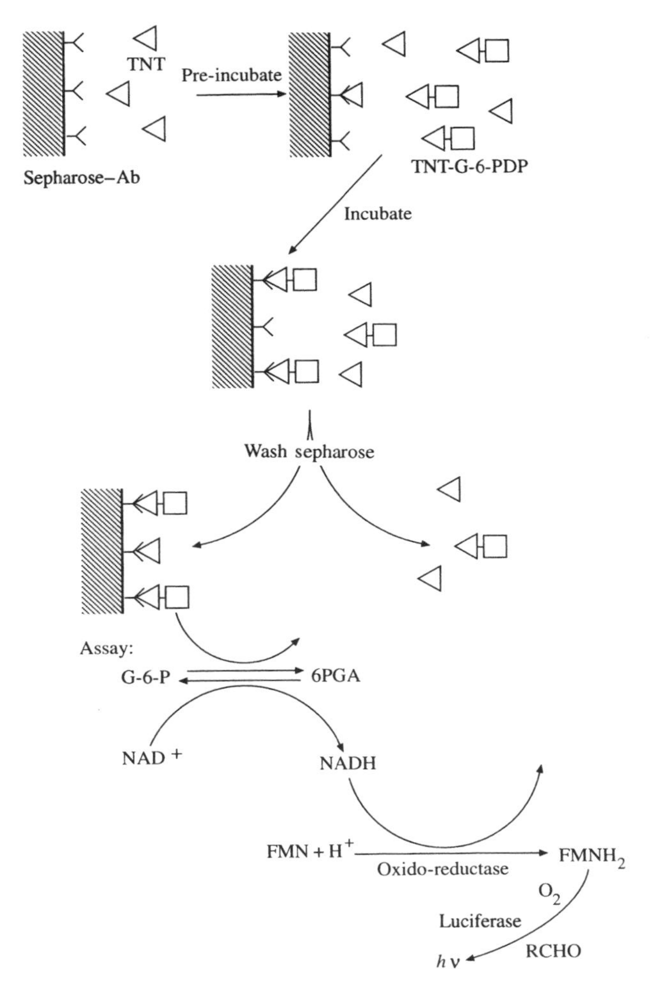

Figure 2.42 Amplified bioluminescent immunoassay of trinitrotoluene (TNT). From Hall, E. A. H., *Biosensors*, Copyright 1990. © John Wiley & Sons Limited. Reproduced with permission.

DQ 2.9

What are the advantages of luminescence methods when compared with other photometric methods?

Answer

Most photometric methods involve the comparison of light absorbed (or emitted) in the presence of the analyte with the corresponding light in the absence of the analyte, which will not usually be zero. However luminescence (chemiluminescence or bioluminescence) is measured against a background of complete absence of light. This means that much more sensitive detectors can be used and much lower levels of light can be detected. Hence the detection limits can be very much lower – as in the example of TNT described above.

SAQ 2.18

What is 'chemiluminescence'?

SAQ 2.19

Mention two sources of bioluminescence.

2.7.6 Optical Transducers

Optical fibres are the key elements required to convert a photometric assay into a sensor. They have really only been developed over the past 15–20 years. Such components could be regarded as 'light conductors' or 'light wires'. Just as metal electrical wires will conduct electricity and bring power and electrical information to the point of need, often over very long distances, so optical fibres will do the same for light. Indeed, optical fibres are even now replacing electrical wires for telephone transmission. Optical fibres behave as *waveguides* for light. The original fibres were made of glass, but polymeric materials are now used, as they are much cheaper than glass and the metal wires used for electricity.

The light waves are propagated along the fibre by *total internal reflection* (TIR), as illustrated in Figure 2.43. Total internal reflection depends on the angle of incidence and the refractive indices of the media, as given by the following equation (see Figure 2.44):

$$\sin\theta/\sin\phi = n_2/n_1 = n \text{ (Snell's law)}$$

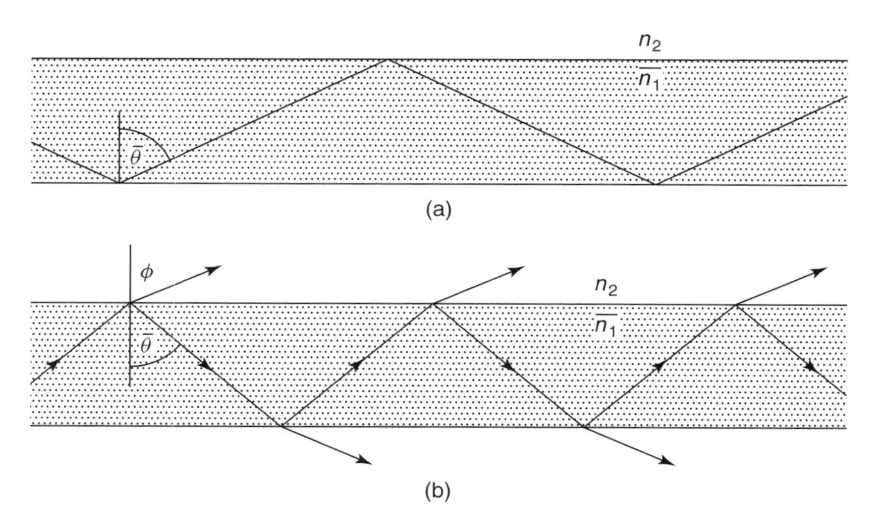

Figure 2.43 Illustration of total internal reflection in a single-core optical fibre, showing the trajectories for (a) bound and (b) refracting rays.

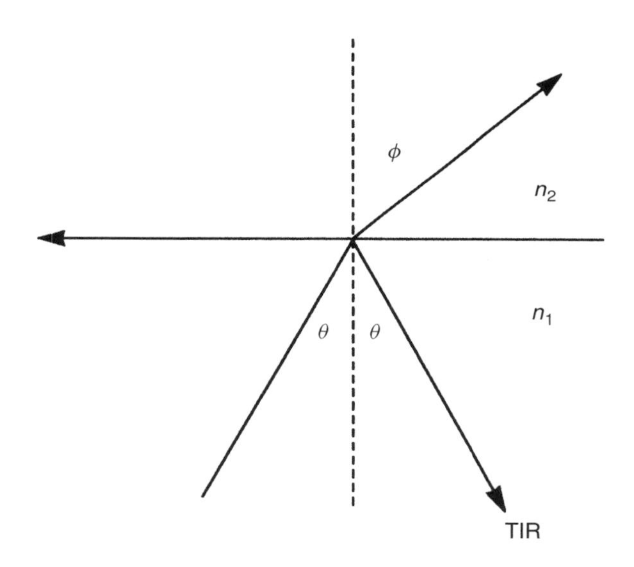

Figure 2.44 Illustration of Snell's law and total internal reflection. From Eggins, B. R., *Biosensors: An Introduction*, Copyright 1996. © John Wiley & Sons Limited. Reproduced with permission.

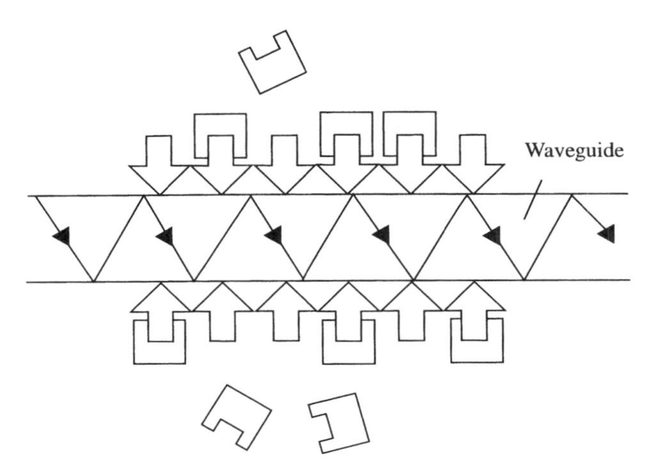

Figure 2.45 Schematic of an optical sensor with evanescent field monitoring. From Hall, E. A. H., *Biosensors*, Copyright 1990. © John Wiley & Sons Limited. Reproduced with permission.

When $\phi = \pi/2$ and $\theta = \theta_c$, then $\theta_c = n_2/n_1$, where θ_c is the critical angle for TIR. Thus, if $\sin\theta > n_2/n_1$, total internal reflection will occur, while if $\sin\theta < n_2/n_1$ we will have refraction and reflection.

These principles need to be applied to a plane harmonic wave and to a cylindrical wave. One must consider the effect of change of phase angle and of bends in the fibre.

Waveguides in sensors are used in two ways, namely the extrinsic or intrinsic modes. In the extrinsic mode, the waveguides simply act to transmit light from the light source to the light collector. In this mode, Beer's law can generally be used. In the intrinsic mode, the light is changed by the measurand – the phase, polarization and intensity can be modulated within the waveguide by a measurand lying within the penetration depth for the evanescent field adjacent to the guide (Figure 2.45).

SAQ 2.20

What is 'total internal reflection'?
Illustrate by the use of a diagram.

SAQ 2.21

How does an optical fibre transmit light?

2.7.7 Device Construction

No reference sensor is needed, unlike electrochemical transducers, for which there must always be a reference electrode. However, the system works better with a 'reference blank', with the light source then being split between the sample and the reference. Detection may be effected at different analytical and reference wavelengths and one then obtains the ratio of the signals at the two wavelengths. This eliminates scatter and source fluctuations.

Waveguides for different forms of light may need to be made of different materials, as follows:

- $\lambda > 450$ nm, plastic (such as polyacrylamide)
- $\lambda > 350$ nm, glass
- $\lambda < 350$ nm, fused silica
- $\lambda > 1000$ nm, germanium crystal

In a photometric sensor, the reagent has to be immobilized so that it can interact with the analyte, probably to form a complex with distinctive optical properties which can then be monitored by the sensor.

2.7.8 Solid-Phase Absorption Label Sensors

The criteria for the application of these are as follows:

- Choice of immobilization support
- Immobilization of indicators, thus retaining activity in the desired rays
- Immobilization of biorecognition molecules with retention of activity
- Cell geometry
- Choice of source and detector components

For a portable device, low-power components are desirable, such as light-emitting diode (LED) sources and photodiode detectors. LEDs have only a limited range of wavelengths at present, with the major ones being shown in Table 2.2.

Table 2.2 Some examples of light sources and detectors and their corresponding wavelengths. From Eggins, B. R., *Biosensors: An Introduction*, Copyright 1996. © John Wiley & Sons Limited. Reproduced with permission

Light source	Wavelength	Detector	Wavelength (nm)
Blue	455–465		
Yellow	580–590	Photodiode	560 (460–750)
Green	560–570	Photodiode	750
Red	635–695	Photodiode	800
Near-infrared	820	Photodiode	850
Infrared	930–950	Photodiode	900 (250–1150)
		Phototransistor	940

The optical properties of an immobilized matrix are dependent on the cell geometry. For example, the indicator may be covalently bonded to polyacrylamide (PAA)-coated micro-spheres, thus resulting in much scattered light being detected.

Figures 2.46–2.48 show three different modes of cell operation.

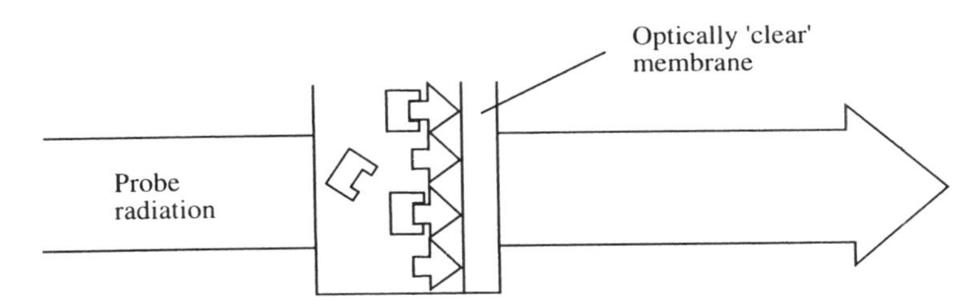

Figure 2.46 Schematic of a solid-state optical sensor. From Hall, E. A. H., *Biosensors*, Copyright 1990. © John Wiley & Sons Limited. Reproduced with permission.

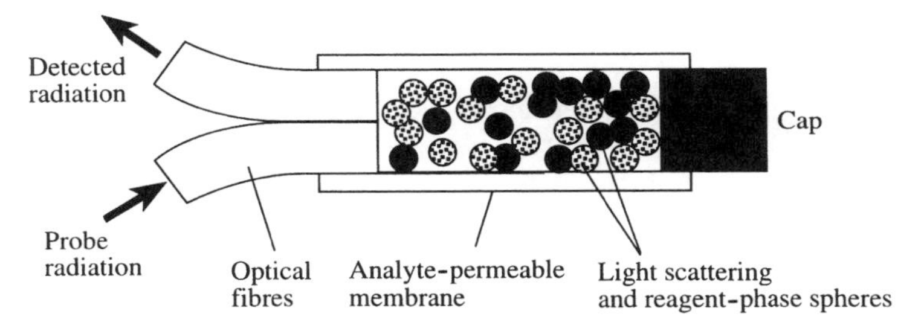

Figure 2.47 Schematic of an optical sensor: light-scattering probe with common source and detection wavelength. From Hall, E. A. H., *Biosensors*, Copyright 1990. © John Wiley & Sons Limited. Reproduced with permission.

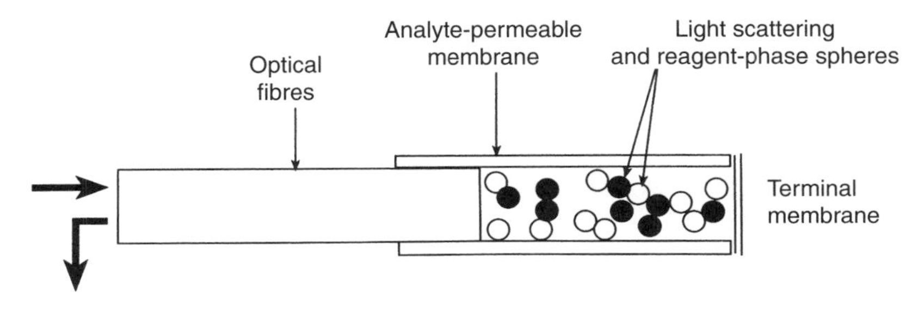

Figure 2.48 Schematic of an optical sensor: light-scattering probe with different source and detection wavelengths. From Hall, E. A. H., *Biosensors*, Copyright 1990. © John Wiley & Sons Limited. Reproduced with permission.

2.7.8.1 Catalysis

The immobilized reagent can be a catalyst for the purpose of converting the analyte into a substance with different optical properties. For example, immobilized alkaline phosphatase catalyses the hydrolysis of *p*-nitrophenyl phosphate to *p*-nitrophenoxide, which has a strong adsorption spectrum.

SAQ 2.22

How can the light source be miniaturized?

2.7.9 Applications

Optical sensors can be used for the measurement of pH, oxygen, carbon dioxide, ammonia and ions, as described later in Chapter 6. Biosensors can be constructed based on these optical sensors, just as they can be based on potentiometric transducers for the same range of analytes.

Summary

The first part of this chapter develops the theory of cell emfs and its application to potentiometry and ion-selective electrodes. An outline background to voltammetry and amperometry then follows (omitting the mathematical derivations). Linear-sweep voltammetry and cyclic voltammetry are discussed, as well as potential step chronoamperometry. There is a brief section about conductivity, including impedance spectroscopy. Miniaturization is discussed within the framework of field-effect transistors, and applications of the latter to direct chemical sensors, ion-selective sensors and enzyme electrodes. Thick- and thin-film electrodes, as well as microelectrodes, are also presented.

The second part gives an introductory survey of the advantages and disadvantages of photometric assays. The optical techniques of UV–visible absorbance and fluorescence spectrophotometrics are presented, and the Beer–Lambert law is given. The adaptation of classical chemiluminescence and bioluminescence to biosensors is described, along with illustrative examples. The use of optical fibres in the fabrication of optical waveguides for sensor development is discussed. The modes of immobilizing photometric reagents and different arrangements of sensor devices are given, followed by a brief summary of major applications.

Further Reading

Albery, W. J. and Cranston, D. H., 'Amperometric enzyme electrode', in *Biosensors: Fundamentals and Applications*, Turner, A. P. F., Karube, I. and Wilson, G. S. (Eds), Oxford University Press, Oxford, UK, 1987, pp. 180–210.

Bartlett, P. N., 'The use of electrochemical methods in the study of modified electrodes', in *Biosensors: Fundamentals and Applications*, Turner, A. P. F., Karube, I. and Wilson, G. S. (Eds), Oxford University Press, Oxford, UK, 1987, pp. 211–246.

Blackburn, G. F., 'Chemically sensitive field effect transistors', in *Biosensors: Fundamentals and Applications*, Turner, A. P. F., Karube, I. and Wilson, G. S. (Eds), Oxford University Press, Oxford, UK, 1987, pp. 481–530.

Carr, R. J. G., Brown, R. G. W., Rarity, J. G. and Clarke, D. J., 'Laser light scattering and related techniques', in *Biosensors: Fundamentals and Applications*, Turner, A. P. F., Karube, I. and Wilson, G. S. (Eds), Oxford University Press, Oxford, UK, 1987, pp. 679–703.

Karube, I., 'Microbiosensors based on silicon technology fabrication', in *Biosensors: Fundamentals and Applications*, Turner, A. P. F., Karube, I. and Wilson, G. S. (Eds), Oxford University Press, Oxford, UK, 1987, pp. 471–480.

Kuan, S. S. and Guillbault, G. G., 'Ion selective electrodes and biosensors based on ISEs', in *Biosensors: Fundamentals and Applications*, Turner, A. P. F., Karube, I. and Wilson, G. S. (Eds), Oxford University Press, Oxford, UK, 1987, pp. 135–152.

McCapra, F., 'Potential applications of bioluminescence and chemiluminescence in biosensors', in *Biosensors: Fundamentals and Applications*, Turner, A. P. F., Karube, I. and Wilson, G. S. (Eds), Oxford University Press, Oxford, UK, 1987, pp. 617–638.

Schultz, J. S., 'Design of fibre optic biosensors based on bioreceptors', in *Biosensors: Fundamentals and Applications*, Turner, A. P. F., Karube, I. and Wilson, G. S. (Eds), Oxford University Press, Oxford, UK, 1987, pp. 639–655.

Seitz, W. R., 'Optical sensors based on immobilised reagents', in *Biosensors: Fundamentals and Applications*, Turner, A. P. F., Karube, I. and Wilson, G. S. (Eds), Oxford University Press, Oxford, UK, 1987, pp. 599–617.

Sutherland, R. M., and Dahne, C., 'IRS devices for optical immunoassays', in *Biosensors: Fundamentals and Applications*, Turner, A. P. F., Karube, I. and Wilson, G. S. (Eds), Oxford University Press, Oxford, UK, 1987, pp. 655–679.

Chapter 3

Sensing Elements

Learning Objectives

- To understand the principles governing selective elements for ions.
- To understand the principles of operation of ion-selective electrodes, and their advantages and limitations.
- To understand the effect of interferences on ion-selective electrodes and how to counteract them.
- To know how to use the Nicholskii–Eisenman equation.
- To be aware of the use and limitations of conductance measurements as selective elements.
- To know of various ways in which electrodes may be modified to make them more selective, such as the use of carbon paste.
- To appreciate how the modification of electrodes by using conducting, ion-exchange and redox polymers can contribute to ionic recognition.
- To learn the principles of selective complexation by using particular reagents.
- To encounter three examples in which kinetic control can be used to improve selectivity.
- To learn about examples of molecules which complex specifically by the 'fitting' of the sample molecule into a space in a complexing molecule of similar size.
- To obtain an outline of the main spectroscopic methods used for molecular recognition.
- To know the different biological materials that can be used as selective agents in biosensors, such as enzymes, including those derived from micro-organisms, organic tissues or mitochondria, antibodies, nucleic acids and receptors.

- To be able to describe the mode of action of each material and to discuss the merits and modes of application of each of them.
- To know the five main methods of immobilizing biological materials on to transducers.
- To understand the relative utility of each method and the situations for which they are most useful.

3.1 Introduction

This chapter is concerned with various ways in which a sensor can recognize an analyte. This recognition should ideally be *specific* for that analyte alone, although sometimes it is just *selective*, i.e. it responds to the required analyte more than to other species. In this case, there may be interferences from other species if they are present in too high a concentration.

We can look at a number of modes of sensing, including the following:

- Ionic
- Molecular
- Biological

At the end of this chapter is a section on methods of attaching the sensor element to the transducer (immobilization).

3.2 Ionic Recognition

3.2.1 Ion-Selective Electrodes – Introduction

Ion-selective electrodes (ISEs) have been described earlier in Chapter 2. These are based on the principle of the emf of a concentration cell. Thus, they are potentiometric devices in which the change in emf is proportional to the logarithm of the analyte concentration. The selectivity is provided by the membrane, which separates the analyte solution from the internal reference solution.

3.2.2 Interferences

Ion-selective electrodes should *not* be called ion-specific electrodes. They respond to one ion more than to others, although there is often a small response to unwanted ions. This is known as *interference*. For example, a fluoride ISE responds to hydroxide at one tenth of the response level (to fluoride) for equal concentrations of the ions. The level of interference is measured by the selectivity coefficient, while the measure of interference is given by the Nicolskii–Eisenman equation (sometimes just called the Nicolskii equation). This is stated as follows:

$$E = \text{constant} + RT/nF \ln (a_i + k_{ij}a_j^{n/z}) \qquad (3.1)$$

where n and z are, respectively, the charges on the primary ion of activity a_i and the interfering ion of activity a_j, and k_{ij} is the selectivity coefficient. Tables of selectivity coefficients are available, with some of these listed below in Table 4.1.

If we consider the fluoride ISE, we see that the selectivity coefficient for hydroxide is 0.1. This means that for equal concentration of fluoride and hydroxide, the response of the fluoride ISE in an hydroxide system is 0.1 times that of the ISE in a fluoride system.

It is necessary to avoid or eliminate the interferants if one wishes to obtain accurate measurements. The selectivity coefficient indicates the relative level of interference, so that we can judge whether special treatment is required to remove it. The fluoride ISE illustrates several interference problems. As well as the hydroxide interference, aluminium and iron(III) can interfere by complexing with the fluoride.

Hydroxide is simply removed by buffering the solution to a slightly acid level. If we set the level to pH 5, it means that the pOH is 9, i.e. the hydroxide ion concentration is 10^{-9} M. This is well below the detection limit of 10^{-6} for fluoride. Both aluminium and iron(III) can be eliminated by adding a more powerful complexing agent, e.g. citrate, which will bind more strongly than fluoride with both aluminium and iron(III). Thus, for the analysis of fluoride by using an ISE, one employs an ionic-strength adjustment buffer (ISAB) containing sodium chloride (1 M) and acetic acid (1 M), adjusted to pH 5.5 with sodium hydroxide and citric acid (10^{-3} M).

In practice, in a real solution there may be several interfering ions and it may be desirable to determine their activities. A relatively new approach to this 'multi-ion problem' is to set up a matrix of several ion-selective electrodes – selective for each of the ions in the solution. The response from each electrode is fed into a computer and the matrix problem is them solved to give the activities of all of the ions measured, taking into account the mutual interferences between each of the ions in pairs. This is an application of *neural network analysis* and has been used to determine, for example, mixtures of potassium, calcium, copper, chloride and nitrate ions. This technique can also be used with conductance measurements.

Other problems with ion-selective electrodes may be due to the effect of matrix components such as proteins or lipids, which can produce gels of high molecular weight, which may lead to deterioration of the selective membrane, thus reducing the working life of the sensor.

SAQ 3.1

What is the purpose of adding an ionic-strength adjustment buffer to each sample and standard?

(a) To ensure that all solutions have the same total ionic strength.
(b) To adjust the ionic strength to 1 M.
(c) To allow emf measurements to be related directly to the activity of the ion being measured.

DQ 3.1

What is the maximum tolerated concentration of silver ions for a 10% error when using a potassium ISE to measure potassium ions in the range 10^{-3} to 10^{-4} M? The selectivity coefficient for silver ions in the presence of potassium ions is 1.00×10^{-4}.

Answer

The maximum allowable error is 10% on 10^{-4} M, i.e. 10^{-5} M.

The selectivity coefficient tells us that 1 M K^+ will give the same response as 10^{-4} M Ag^+. Therefore, the concentration of K^+ which will give a response similar to 10^{-5} M Ag^+ is 0.1 M.

3.2.3 Conducting Devices

Ions in solution cause the latter to conduct an electric current. There is a certain amount of selectivity in this process, which is based on the following principles. The conductance of an electrolyte solution is the quantity measured, where conductance is the reciprocal of resistance. It is the usual practice to normalize this parameter (with respect to the dimensions of the conductivity cell) to give the conductivity. This is often carried out automatically in modern instruments. The value of the conductance depends on the following factors:

- The degree of ionization – this distinguishes between strong electrolytes, such as salts, and weak electrolytes, such as organic acids and bases
- The charge on the ion
- The number of ions in solutions – (roughly) the concentration of the ions
- The mobility of each ion – this is related to the size of the ion, with a small ion having a larger mobility than a large ion and thus contributing to a higher conductivity

Conductivity has not been used in sensor development as extensively as it might, as realistically the system requires another selective element to be coupled with it. However, conducting devices are used as detectors in ion chromatography.

3.2.4 Modified Electrodes and Screen-Printed Electrodes

3.2.4.1 Introduction

Electrodes themselves do not possess selectivity, except that which is due to variations of the imposed potential. However, by modifying the electrode in various ways (as described below), degrees of selectivity may be introduced into the electrode itself. The simple carbon paste electrode is very easy to modify. The incorporation of polymers of various sorts, either on the electrode surface or in the matrix of the electrode itself, provides a variety of more sophisticated ways of making the electrode more selective.

3.2.4.2 Modified Carbon Paste Electrodes

One of the simplest types of (modified) electrode is the modified *carbon paste electrode* (CPE). This was introduced in the 1960s for studying oxidation reactions where mercury is not a suitable electrode material. It consists of a simple mixture of graphite power with 'Nujol' to form a stiff paste, which is then placed into an electrode holder, as shown in Figure 3.1.

It is simplicity itself to mix a modifying component in with the paste. Such a component could be itself electroactive, as with ferrocene, used in the glucose biosensor, or maybe a complexing agent, which can extract an electroactive analyte into the surface of the paste.

Another biosensor application is the 'bananatrode' in which banana pulp is mixed with the paste and used to determine dopamine and other catechols (see Chapters 2 and 8 for further details).

Another use of CPEs involves mixing a complexing agent with the paste, which can then extract an electroactive analyte into the surface of the paste. Thus, 2,9-dimethyl-1,10-phenanthroline has been used to complex copper(I) and to determine the latter by voltammetric oxidation to a detection limit of 3×10^{-7} M, with no interference from Cu(II), Zn(II), Co(II), Pb(II), Ni(II) or Fe(II), and very little from Ag(I) (see Section 8.2 below). A similar method has been developed for the determination of Ag(I), employing a different phenanthroline.

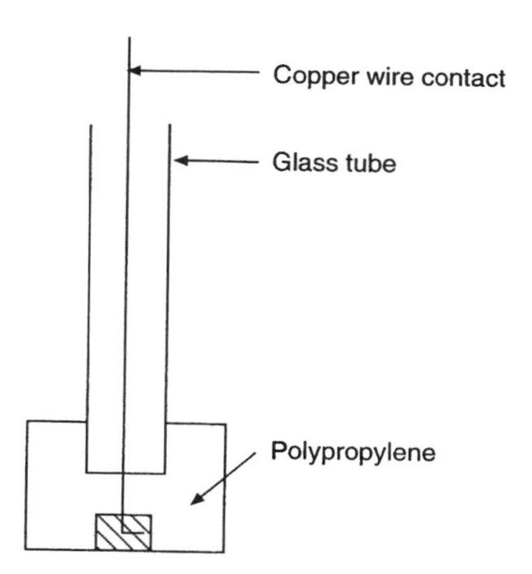

Figure 3.1 Schematic of a carbon paste electrode. From Eggins, B. R., *Biosensors: An Introduction*, Copyright 1996. © John Wiley & Sons Limited. Reproduced with permission.

3.2.4.3 Polymer Electrodes

A great deal of research has been carried out in recent years into modifying electrode surfaces by coating them with different types of polymers. These have been mainly of three types, i.e. conducting polymers, ion-exchange polymers and redox polymers.

Often, chemical groups are attached to these coatings (or incorporated into their structures) in order to introduce particular electrochemical effects. The applications have mainly been concerned with voltammetric processes, although they have also been used in both potentiometric and conductometric applications.

3.2.4.4 Electronically Conducting Polymers

The most studied conducting polymers have been polyacetylene, polypyrrole, polyaniline and polythiophene. These are easily prepared by electrochemically oxidizing the substrate on the electrode surface. The solvent used, and more particularly the counter anion in the solution, have a major effect on the properties of the polymer, and in particular on its selectivity characteristics for use in sensors. We shall give most attention to polypyrrole as it is the most versatile and the most studied of such materials. However, similar principles also apply to other types of conducting polymers.

A solution of pyrrole in aqueous 0.1 M KCl can be oxidized at +0.8 V (vs. SCE) at a platinum or glassy carbon electrode to form a layer of polypyrrole on the electrode surface, according to the sequence shown in Figure 3.2. Chemical methods of preparation can be used, such as oxidation by halogens, SbF_5, AsF_5 or $FeCl_3$. The electrochemical method can be varied by changing the solvent and, more interestingly, the electrolyte. The most important electrochemical property of these conducting polymers is their redox switching behaviour. The polymers are electroactive and at certain potentials, as illustrated by the cyclic voltammogram shown in Figure 3.3, reduction can occur from the electronically

Figure 3.2 Oxidative polymerization of pyrrole. From Eggins, B. R., *Biosensors: An Introduction*, Copyright 1996. © John Wiley & Sons Limited. Reproduced with permission.

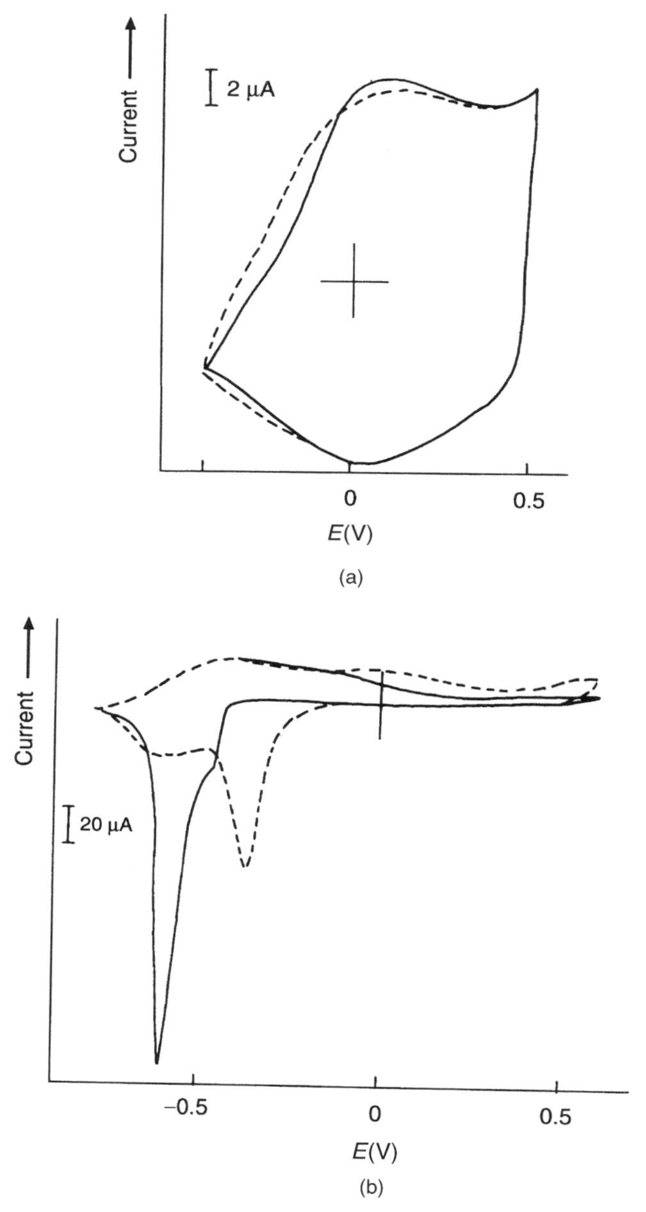

Figure 3.3 Cyclic voltammograms of polypyrrole with different counter anions: (a) in chloride; (b) in dodecylbenzenesulfate. From Lyons, M. E. G., Lyons, C. H., Fitzgerald, C. and Banon, T., *Analyst*, **118**, 361–369 (1993). Reproduced with permission of The Royal Society of Chemistry.

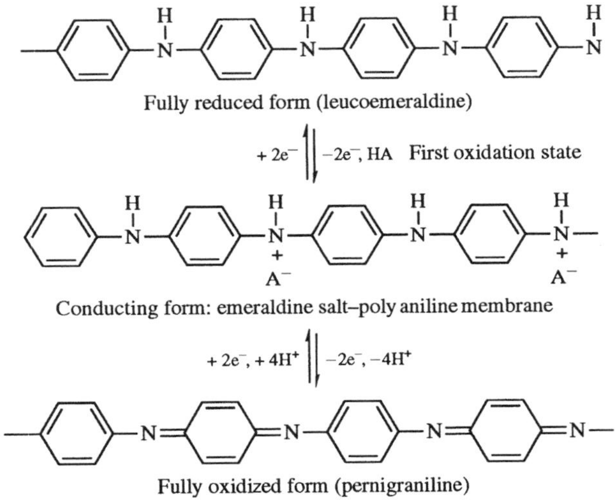

Fully reduced form (leucoemeraldine)

$+2e^-$ / $-2e^-$, HA First oxidation state

Conducting form: emeraldine salt–poly aniline membrane

$+2e^-, +4H^+$ / $-2e^-, -4H^+$

Fully oxidized form (pernigraniline)

Figure 3.4 Oxidation of aniline to polyaniline showing the various switching modes. From Teasdale, P. R. and Wallace, G. G., *Analyst*, **118**, 329–334 (1993). Reproduced with permission of The Royal Society of Chemistry.

conducting state to the non-electronically conducting form (although the material still remains ionically conducting):

$$\text{PPY} + \text{A}^-_{(\text{soln})} \rightleftharpoons \text{PPY}^+ + \text{A}^-_{(\text{poly})}$$

Polyaniline has three different oxidation states, and hence three switching modes, as shown in Figure 3.4. Such changes have a dramatic effect on the electrical resistance of the material, thus resulting in large changes in its molecular recognition capabilities.

A wide range of types of molecular interaction contributes to variations in molecular recognition, with these being principally the following:

- ion–ion
- ion–dipole
- dipole–dipole (hydrogen bonding)
- ion–induction
- dipole–induction
- dispersion

Analyte recognition is generally by either (a) chelation and complexation, (b) enzyme–substrate interaction, or (c) antibody–antigen interaction. These can all be accomplished by incorporation of the appropriate group into the polymer matrix, such as ethylenediaminetetraacetic acid (EDTA) in polypyrrole, or by incorporation of biorecognition elements such as glucose oxidase or urease into the polymer. Work has also been carried out with polynucleotides and antibodies.

3.2.4.5 Ion-Exchange Polymers

Conducting polymers are inherently *ion-exchange materials*, and the nature of the counter-ion can greatly modify the relative exchange properties of the polymer. Thus, with polypyrrole chloride, the order of exchange is $Br^- > SCN^- > SO_4^{2-} > I^- > CrO_4^{2-}$, while with polypyrrole perchlorate, the order is $SCN^- > Br^- > I^- > SO_4^{2-} > CrO_4^{2-}$.

If one uses a hydrophobic counter-ion such as dodecylsulfate or poly(vinyl sulfonate), the ion-exchange character can be eliminated or modified. The polymer can also be transformed from an anion-exchange polymer into a cation exchanger, as shown in the scheme and the accompanying cyclic voltammogram presented in Figure 3.5.

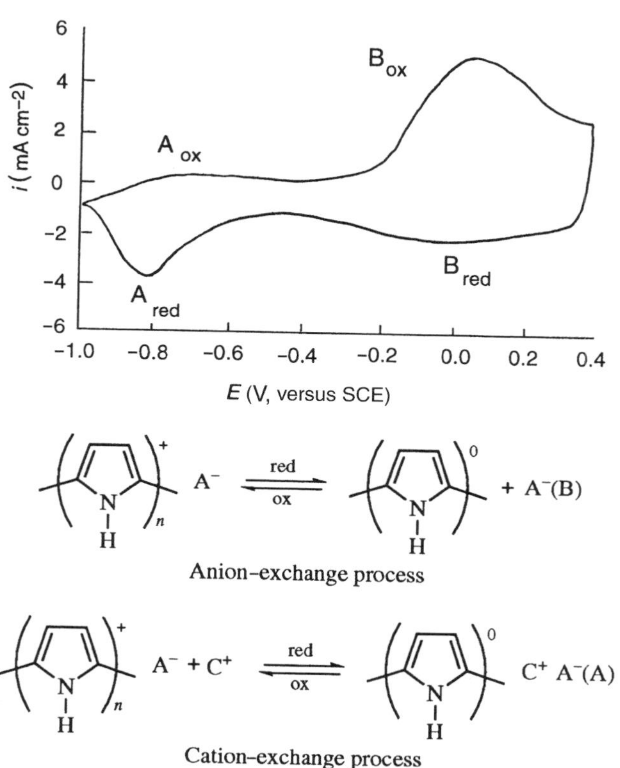

Figure 3.5 Pyrrole-based anion–cation-exchange polymers. The cyclic voltammogram is that of polypyrrole at a microelectrode in 10^{-3} mol l^{-1} KCl. From Lyons, M. E. G., Lyons, C. H., Fitzgerald, C. and Banon, T., *Analyst*, **118**, 361–369 (1993). Reproduced with permission of The Royal Society of Chemistry.

SAQ 3.2

What are the ways in which conducting polymers can act as ion-selective electrodes?

Ionomer-film-modified electrodes represent another type of polymer-coated electrode. An ionomer is a linear- or branched-chain polymer containing covalently attached ionizable groups. Such materials contain <10% of ionizable monomers and are not cross-linked.

They are not themselves electroactive, but show unusually high ion-exchange selectivity for large hydrophobic cations, such as alkylammonium ions, when compared with mono- and divalent inorganic ions. The main types are perfluorosulfonate polymers, such as 'Nafion' ($[CF_3(CF_2)_n]_2{=}CF{-}O{-}CF_2{-}CF(CF_3){-}O{-}(CF_2)_2{-}SO_3H$) (du Pont) and 'PFASA' ($[CF_3(CF_2)_n]_2{=}CF{-}O{-}(CF_2)_2{-}SO_3H$) (Dow).

The ionomer film will extract and pre-concentrate large cations from an aqueous phase. The cations which have been studied include electroactive ions such as methyl viologen (MV^{2+}), ferrocenylmethyltrimethylammonium (FA^+), and ruthenium complexes such as $Ru(NH_3)_4{}^{3+}$ and $Ru(byp)_3{}^{2+}$. These can all be detected by voltammetry.

The interaction between the aqueous solution and the film can be expressed by the following equation:

$$O^{n+}(aq) + nNa^+(film) \longrightarrow O^{n+}(film) + nNa^+(aq)$$

This reaction can be quantified by the distribution coefficient, k_D, between water and the film, and $K_{O/Na}$, the equilibrium constant for the reaction. The values of k_D and $K_{O/Na}$ for these ions are given in Table 3.1, while some typical values of $K_{O/Na}$ for simple inorganic ions are $Cs^+ = 9.1$ and $Ba^{2+} = 30$.

Table 3.1 Ion-exchange distribution coefficients, k_D, and selectivity coefficients, $K_{O/Na}$. From Espensheid, M. W., Ghatak-Roy, A. R., Moore, III, R. B., Penner, R. M., Szentirmay, M. N. and Martin, C. R., *J. Chem. Soc., Faraday Trans. 1*, **82**, 1051–1070 (1980)

Cation	k_D	$K_{O/Na}$
MV^{2+} [a]	7.9×10^5	1.5×10^4
FA^+ [b]	1.1×10^6	7.3×10^4
$Ru(NH_3)_6{}^{2+}$	2.5×10^6	3.7×10^4
$Ru(bpy)_3{}^{2+}$	2.1×10^7	5.7×10^6
$Ru(NH_3)_6{}^{2+}$	2.6×10^4	7.4×10^2

[a] Methyl viologen.
[b] Ferrocenylmethyltrimethylammonium.

> **SAQ 3.3**
>
> How can the anion affect the properties of a conducting polymer?

From the analytical point of view, the procedure is similar to that used in anodic stripping voltammetry (ASV). As can be seen in Figure 3.6, concentrations of 10^{-8} M can easily be detected with cyclic voltammetry (CV) currents of about 2–20 μA. Such a technique can be termed *ion-exchange voltammetry*.

In order to establish a viable system, a number of factors need to be considered, as follows:

(i) Film mass transport dynamics, which govern the speed of film/solution equilibration.

(ii) Ion-exchange selectivity, which governs the detection limits.

(iii) Regeneration of the film for a new run.

(iv) Selectivity between analytes. Ionomers have only a very rudimentary selectivity, although they are better than the bare electrode. This factor needs to be further studied. Approaches may be to build selectivity into the membrane, as with the chemically selective membranes used in potentiometry.

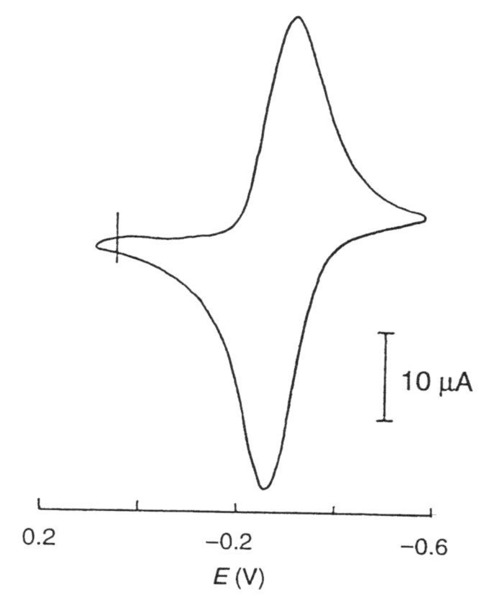

Figure 3.6 Cyclic voltammogram of a 'Nafion'-coated electrode in equilibrium with 2.78×10^{-8} M $Ru(NH_3)_6^{3+}$. From Espensheid, M. W., Ghatak-Roy, A. R., Moove, III, R. B., Penner, R. M., Szentirmay, M. N. and Martin, C. R., *J. Chem. Soc., Faraday Trans. 1*, **82**, 1051–1070 (1980). Reproduced with permission of The Royal Society of Chemistry.

An interesting application of this technique is the *in vivo* determination of neurotransmitter substances such as dopamine in the brains of rats, in the presence of potentially interfering substances such as ascorbate. The former are protonated at physiological pH levels and will be extracted by the 'Nafion' film, whereas the ascorbate remains as the anion and is thus rejected.

SAQ 3.4

What is meant by the term 'ion-exchange voltammetry'?

3.2.4.6 Redox Polymers

A number of modified electrodes have been made with a redox group attached to the polymer coating. A popular approach has been to polymerize 4-vinylpyridine on the surface. This will then selectively coordinate with transition metal ions. Ruthenium and osmium ions have been particularly studied, as the resulting redox polymers are effective redox catalysts for other analytes.

A similar approach has been used with polypyrroles, poly-*N*-methylenepyrroles and polythiophenes, using mainly covalently attached quinones as the redox group, as illustrated by the example shown in Figure 3.7. Most of this work has been carried out using chlorinated benzoquinones and naphthoquinones, and a sulfonated anthraquinone, in order to keep the redox potential as low as possible.

Another approach has been to polymerize amino acids containing built-in redox groups. Such polymerizations occur readily at moderate pH levels. They have been developed with 4-nitrobenzoyl groups attached to poly-L-lysine, poly-L-ornithine and poly-L-glutamate. With an imidazole group attached to poly-L-ethylglutamate, an iron porphyrin group was attached as the redox group via a trimethyleneamino group to make a cytochrome *c* analogue.

SAQ 3.5

How can electrodes be modified with redox groups?

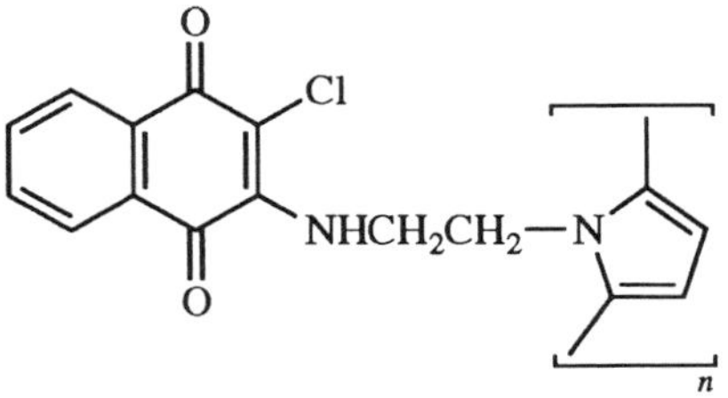

Figure 3.7 A redox polymer consisting of a quinone attached to polypyrrole. From Eggins, B. R., *Biosensors: An Introduction*, Copyright 1996. © John Wiley & Sons Limited. Reproduced with permission.

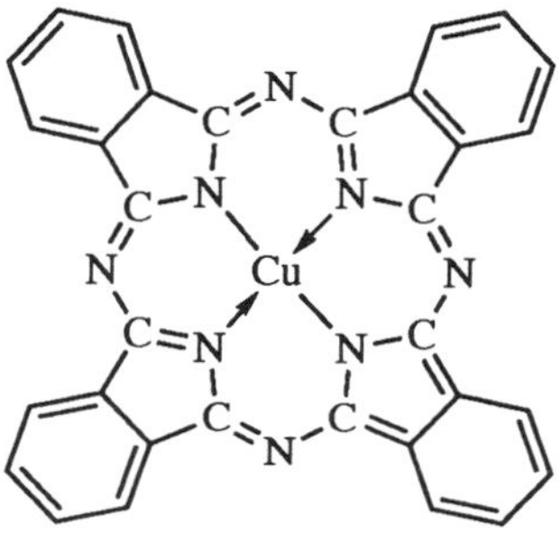

Figure 3.8 Structure of copper phthalocyanine. From Eggins, B. R., *Biosensors: An Introduction*, Copyright 1996. © John Wiley & Sons Limited. Reproduced with permission.

3.2.4.7 Phthalocyanines

Phthalocyanines (Figure 3.8) are weak semiconductors, but this conductivity can be greatly increased by complexation with certain electron-acceptor molecules such as NO_2, which makes them potential sensor materials for such molecules.

These materials have been used in the conductometric mode in sensors for gases such as NO_2, NO, Cl_2, F_2 and BF_3. Changing the structure of the molecule, in particular the metal, or the substituents on the benzene rings, can vary the sensitivity and selectivity. Unfortunately, they often have to be operated at elevated temperatures such as 80–150°C, and have slow recovery times, as they form donor–acceptor complexes with the substrate molecules. Lead phthalocyanine will detect down to 1 ppb of NO_2 in air. Electron-donor molecules, such as NH_3 and H_2S, interfere to some extent, although hydrocarbons do not.

The redox properties of phthalocyanines can be used to provide an electrocatalytic sensor membrane for the voltammetric/amperometric measurement of species such as cysteines, reduced glutathiones, oxalic acid and hydrazines, either in a carbon paste or in a screen-printed carbon electrode.

DQ 3.2

Consider the ways in which phthalocyanines can be used to make electrodes selective.

Answer

They can be used in the conductive mode to detect gases such as NO and NO_2. Changing the metal can change the selectivity of the molecule.

They can also be employed for the voltammetric measurement of cysteines, etc. in carbon paste or screen-printed electrodes.

3.3 Molecular Recognition – Chemical Recognition Agents

3.3.1 Thermodynamic – Complex Formation

Thermodynamics controls the equilibrium constant between the reacting species and the products. If this is high for one analyte–ligand complex, but low for complexes of other analytes with the same ligand, it can form the basis of a selective method:

$$M + nL = ML_n$$

where:

$$K = [ML_n]/[M][L]^n$$

This principle forms the basis for a vast range of spectroscopic analyses in the UV–visible region, and can also be adapted for sensor technology. The reagent must complex selectively with the analyte and must respond to a particular transducer (usually optical or electrochemical). The optical response is usually via an absorption or fluorescence change to the analyte. Ideally, the reagent shows complete reversibility on removal of the analyte. It is usually necessary to place the ligand in a membrane, one side of which is connected to the transducer, with the other side in contact with the analyte solution.

PVC membranes have been used successfully for a range of such sensors, including ion-selective electrodes (see Table 3.2). These make use of the potentiometric transducer. For example, a nitrate sensor can use the ionophore, tri(*n*-decyl)methylammonium nitrate (1%), with a plasticizer, dibutyl phthalate (66%), and PVC (33%) (as shown in Figure 3.9). Further examples are shown in Table 3.2.

By incorporating the ligand material into the electrode surface, for example, by mixing with carbon paste, one can make amperometric sensors. The complex must be more readily reducible (oxidizable) than the uncomplexed analyte. An

Table 3.2 Some examples of PVC-based sensor membranes

Species	Ionophore
H^+	Tri-*n*-dodecylamine
Li^+	12-crown-4
$NH_4{}^+$	Nonactin
Ba^{2+}	Nonylphenoxy poly(oxyethylene) ethanol
Zn^{2+}	Zinc di-*n*-octylphenyl phosphate
Anionic detergents	Cetyltrimethylammonium dodecylbenzenesulfonate
Paraquat	Paraquat bis(tetraphenyl borate)
Promethazine	Promethazine tetraphenyl borate
Ephedrine	Ephedrine tetraphenyl borate

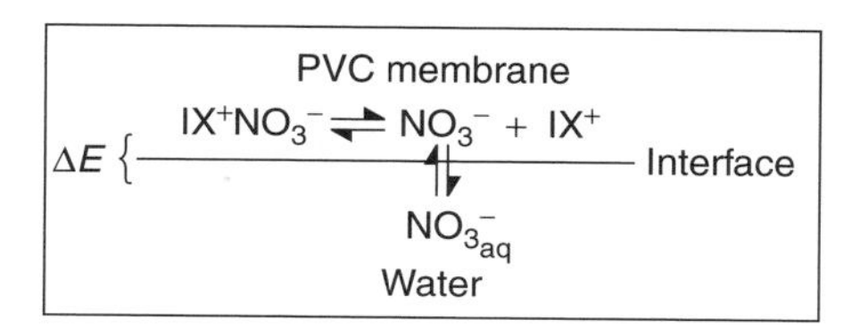

Figure 3.9 Schematic of the establishment of the interfacial potential for a nitrate/ion-exchanger membrane. © R. W. Catterall 1997. Reprinted from **Chemical Sensors** by R. W. Catterall (1997), by permission of Oxford University Press.

example is the analysis of iron(II) by using the ligand 2,2′-bipyridyl in a poly(vinyl pyridine) (PVP) membrane. The complexed iron(II) is then oxidized by using linear-sweep voltammetry. The complexation effectively pre-concentrates the analyte, as well as improving the selectivity.

A range of optical sensors (optodes) uses this principle. A pH indicator changes colour over a certain pH range. Such an indicator may be incorporated into a membrane such as PVC or cellulose acetate. If the indicator has a positive charge in the acid form and is red, and then is neutral in the base form, which is blue, it can act as a pH indicator over the range covered by the colour change from red to blue. This colour change will be detected by an appropriate colorimeter or spectrometer. Electroneutrality in the membrane is achieved by employing a large lipophilic anion such as tetraphenylborate (TPB) (Figure 3.10).

The membrane colour is controlled by the equilibrium reactions shown in this figure, which are controlled by the analyte pH. The absorbance of the membrane is dependent on the ratio of the activities of the hydrogen ion to the sodium ion (a_{H^+/Na^+}). Either a colour-sensitive indicator such as phenol red, or a fluorescent indicator such as 8-hydroxy-1,3,6-pyrenetrisulfonic acid (HPTS), may be used.

Similar principles apply to the sensing of metal ions such as zinc, for which 4-(2-pyridylazo)resorcinol (PAR) is immobilized in a PVC membrane, as shown in Figure 3.11. The intensity of colour of the complex is related to the amount of zinc, although it is also related to the square of the hydrogen ion activity. Sample solutions are made by covalently binding the zinc to cellulose and these solutions must be buffered, e.g. at pH 4.8. Morin (4,8,3,5,7,2′,4′-pentahydroxyflavone) forms a fluorescent complex with Al and Be, and optode membranes can be made by covalently binding this compound to cellulose.

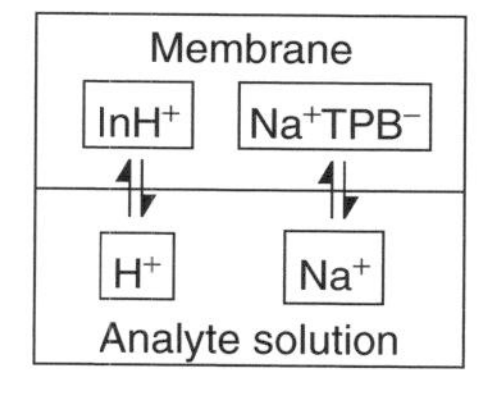

Figure 3.10 Schematic of the mechanism involved in the operation of a pH optode. © R. W. Catterall 1997. Reprinted from **Chemical Sensors** by R. W. Catterall (1997), by permission of Oxford University Press.

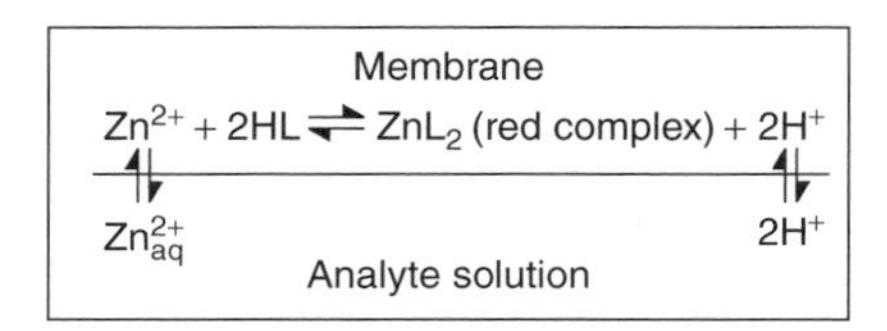

Figure 3.11 Reaction of zinc with a PAR (HL)-based optode membrane. © R. W. Catterall 1997. Reprinted from **Chemical Sensors** by R. W. Catterall (1997), by permission of Oxford University Press.

DQ 3.3

What characterizes the optimal (sensor) reagent used in optical sensors?

Answer

The material should be very sensitive to the analyte, show complete reversibility on removal of the analyte, respond quickly and show a high degree of selectivity for the analyte. It should exhibit photo-stability and not suffer from reagent leaching. For a low-cost sensor, it should also be compatible with low-cost semiconductor light sources, such as LEDs.

3.3.2 Kinetic–Catalytic Effects: Kinetic Selectivity

The rate of a chemical reaction can sometimes be used to control the selectivity. Modified electrodes can be used to lower the activation over-potential and hence effectively speed up the reaction, thus reducing interferences and lowering the limit of detection. An example of this is the analysis of iron (III) at an [Os $(byp)_2(PVP)_{10}Cl]Cl$-coated electrode. Such substances act as an electron shuttle between the electrode (transducer) and the analyte solution. This is similar to the use of mediators in biosensors, which speed up the transfer of electrons between an enzyme and the electrode. The mechanism is as follows:

$$A = B + e^-$$

$$B + Y = A + Z$$

The use of enzymes in biosensors is another example of a kinetic effect. Thus, a reaction can be very fast in the presence of a specific enzyme, while other analytes may not react.

Catalytic gas sensors for flammable gases are based on the controlled combustion of the gas in air over a catalyst such as Pt, Pd, or Rh at ca. 1000°C. A thermistor device, such as a platinum resistance thermometer, detects the heat that is evolved.

A further kinetic effect has been demonstrated in flow-injection analysis (FIA) using an ion-selective electrode. If a sample ion and an interfering ion have the same thermodynamic behaviours, i.e. produce the same potential response, but different kinetic behaviours, this can be used to improve the selectivity. There

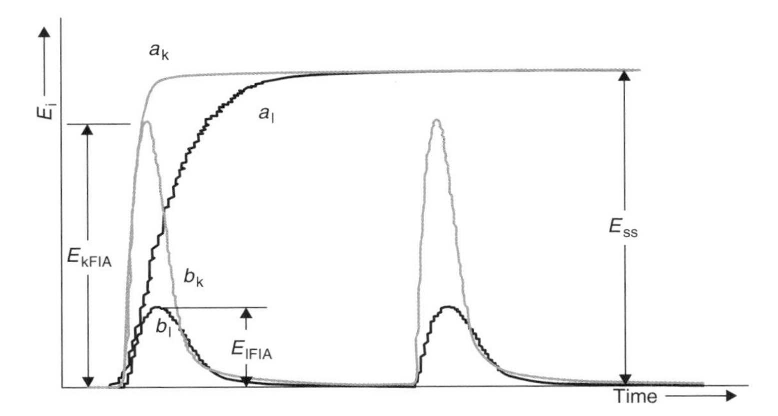

Figure 3.12 Kinetic-enhanced selectivity in ISEs: a_k and a_l, activities of primary and interfering ions, respectively, in the steady state; b_k and b_l, activities of primary and interfering ions, respectively, in flow-injection analysis (FIA). From Diamond, D. (Ed.), *Principles of Chemical and Biological Sensors*, © Wiley, 1998. Reproduced by permission of John Wiley & Sons, Inc.

is a fast exchange across the selective membrane between the primary ions on each side, when compared with the interfering ions. The result is that the sample ions give a much higher response peak than the interfering ions even though they would give equal responses in the steady state (see Figure 3.12)

3.3.3 Molecular Size

Molecular size can be the *basis* for selectivity. In a very simple way, molecular sieves can select between large and small molecules.

Calixarenes (Figure 3.13) are cup-like molecules designed with a specific internal size into which specific molecules can fit. Such materials have been used in sensors for calcium. The dye 'Nile Blue' is incorporated into a poly(vinyl chloride) (PVC) membrane with an anionic counter-ion and a calix[4]arene tetraphosphine oxide calcium-selecting ligand. This is used as an *optode* and selects calcium 'against' sodium. Caesium ions can be trapped in a similar but larger calix[6]arene material.

The antibiotic, valinomycin (Figure 3.14), is a neutral ionophore, which will complex potassium ions selectively because such ions fit nicely into the cavity created within the interior of the valinomycin molecule. This is now the standard method used for the determination of potassium.

Crown ethers and thiacrowns (see Figure 3.15) of varied size can be selective for silver ions (monothiacrown ethers), for potassium (bis-crown ethers), for lithium ions (12-crown-4) and for iron(III) (1,7-dithia-12-crown-4).

The selectivities of sensors made with these materials is similar to those of other ion-selective electrode systems. However, the main advantage is the ability

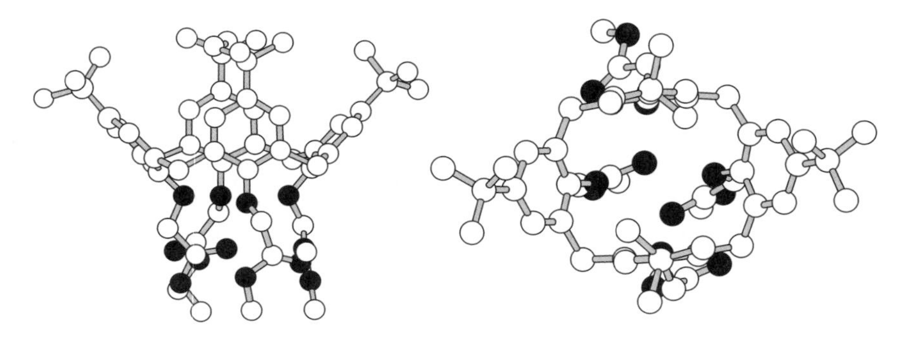

Figure 3.13 Energy-minimized three-dimensional diagrams of a typical calixarene ionophore, i.e. calix[4]:tetramethoxyethyl ester, showing the oxygen atoms that form the negatively charged cavity inside the molecule, which is suitable for ion coordination: ○, carbon; ●, oxygen: hydrogen atoms are not shown for clarity. From Diamond, D. (Ed.), *Principles of Chemical and Biological Sensors*, © Wiley, 1998. Reproduced by permission of John Wiley & Sons, Inc.

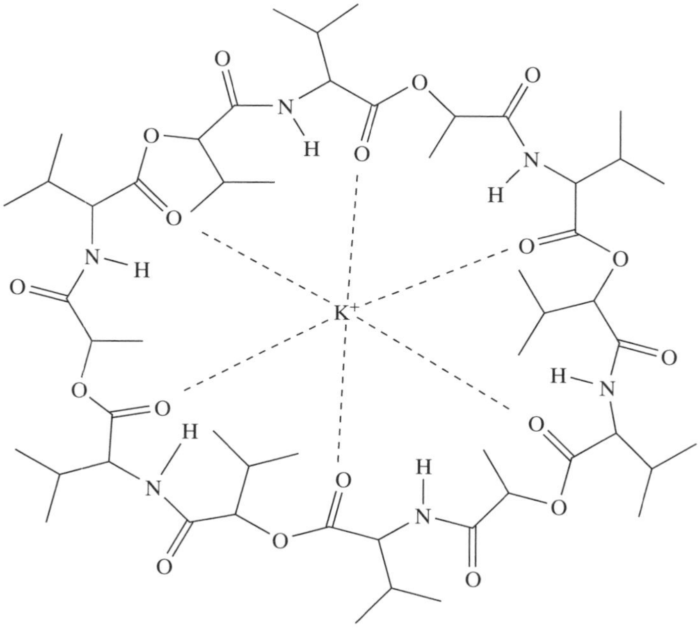

Figure 3.14 The structure of the valinomycin/potassium complex. © R. W. Catterall 1997. Reprinted from **Chemical Sensors** by R. W. Catterall (1997), by permission of Oxford University Press.

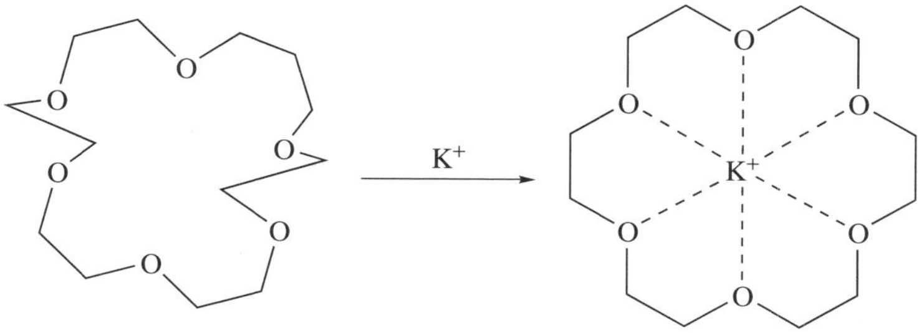

18-crown-6

Complex of 18-crown-6 with K^+

Figure 3.15 Structures of a typical crown ether, 18-crown-6, and its complex with potassium.

to design a membrane which is selective to any given cation. The detection limits are set by the distribution coefficient between the membrane and the sample, while the lifetimes are limited by the leaching of ion exchanger or neutral carrier from the membrane.

SAQ 3.6

How is selectivity imparted by ion-exchange polymer electrodes?

3.4 Molecular Recognition – Spectroscopic Recognition

3.4.1 Introduction

Spectroscopic analysis has been used extensively for many years for the fairly specific identification of the structures of various molecules. The details, theory, practical aspects and applications are covered in other texts referred to at the end of this chapter and in the Bibliography, and so will not be discussed in any detail here. However, for the sake of completion, the principles aspects of molecular spectroscopy will be presented. The internal energy of a molecule can be expressed in the following way:

$$E_{\mathrm{U}} = E_{\mathrm{kin}} + E_{\mathrm{el}} + E_{\mathrm{vib}} + E_{\mathrm{rot}} + E_{\mathrm{nuc}}$$

Theses energies correspond to particular regions of the spectrum and when such radiation falls on to a molecule the latter will absorb the radiation at particular wavelengths so that the molecule goes into a higher energy level.

3.4.2 *Infrared Spectroscopy – Molecular*

The interaction of infrared absorption radiation with organic molecules involves vibrational and rotational energies (E_{vib} and E_{rot}, respectively). This results in a complex pattern of absorption peaks. Some of these can be related to particular interatomic groups in the molecule, such as O–H, C=O, etc. The overall pattern is a unique 'fingerprint' for a particular molecule, and if a match can be obtained, this provides an excellent diagnostic tool.

3.4.3 *Ultraviolet Spectroscopy – Less Selective*

Ultraviolet and visible radiation cause changes in the electronic structure of a molecule (E_{el}). This gives less information than IR spectroscopy, although it can provide useful additional clues, particularly concerning conjugated double-bond structures, including those in aromatic molecules.

3.4.4 *Nuclear Magnetic Resonance Spectroscopy – Needs Interpretation*

Nuclear magnetic resonance (NMR) spectroscopy involves the interaction of certain atomic nuclei with a magnetic field (E_{nuc}). This technique gives fairly specific information about the 'atmosphere' around a particular nucleus. This technique was originally applied mainly to the hydrogen nucleus, ^{1}H, and thus it could distinguish between hydrogen atoms in different 'atmospheres'. A sharp band or pattern would be given by the hydrogens in a CH_3 group, quite distinct from the hydrogen in (say) a –CH=C group. Interpretation of NMR spectra requires some skill to account for interactions between hydrogen nuclei on neighbouring carbon atoms. However, it is a very powerful analytical tool. It is now fairly common to obtain information from the magnetic resonance behaviour of other nuclei, such as ^{19}F, ^{13}C, and several others.

3.4.5 *Mass Spectrometry*

In mass spectrometry (MS), a stream of vaporized molecules is ionized in an electron beam. A detector then separates the fragments with different mass/charge ratios. There is usually a parent ion corresponding to the mass of the whole molecule, although there are also ions present from the breakdown of the molecule into smaller fragments. Considerable theory now exists to enable experts to reconstruct the original molecule from the fragmentation pattern. The latter can be very specific, although caution is needed in interpretation. However, it is more common to use the pattern as a 'fingerprint', as in IR spectroscopy, and to attempt to match the observed pattern to that of authentic specimens. Libraries of MS data exist which can be matched by a computer with the data obtained for an 'unknown' sample. An indication is often given of the probability of the match being good.

DQ 3.4

Summarize how spectroscopic methods can act as 'selective elements' in sensor applications.

Answer

In combination, the four spectroscopic methods described above can usually identify the full structures of most substances, particularly organic compounds. A match of the infrared spectrum with that of a standard substance has long been a definitive way of identification. (Mass spectra can be used in a similar way.) There is a sufficient range of energy bands in IR spectra that virtually any compound can produce a unique pattern. If the analyst does not know what the substance is likely to be, then a combination of methods will be needed, i.e. the IR spectrum to indicate which organic groups are present, NMR data to show how these groups are related to each other, plus the mass spectrum to provide at least the formula mass of the parent ion and hence of the compound. The UV–visible spectrum can also assist in showing which groups are present, and if they are conjugated to each other.

3.5 Molecular Recognition – Biological Recognition Agents

3.5.1 Introduction

Biological systems provide the major selective elements used in biosensors. These must be substances that can attach themselves to one particular substrate, but not to others. Four main groups of materials can achieve this, namely:

- Enzymes
- Antibodies
- Nucleic acids
- Receptors

The biological elements most regularly employed are enzymes. These may be used in a purified form, or may be present in micro-organisms or in slices of intact tissue. They act as biological catalysts for particular reactions and can bind themselves to the specific substrate. This catalytic action is made use of in the biosensor.

Antibodies have a different mode of action. They will bind specifically with the corresponding antigen, removing it from the sphere of activity, but they have no catalytic effect. Despite this, they are capable of developing ultra-high sensitivities in biosensors. Considerable ingenuity is often needed to 'involve them' with the substrate and to provide a signal for the transducer to measure.

Nucleic acids have been much less used so far. They operate selectively because of their base-pairing characteristics. Such compounds have great potential utility in identifying genetic disorders, particularly in children.

Inside the lipid bilayer plasma membrane surrounding a cell are proteins that traverse the full breadth of the membrane and which possess molecular recognition properties. These are known as receptors. They are difficult to isolate, but will bind solutes with a degree of affinity and specificity matching those of antibodies.

3.5.2 Enzymes

An enzyme is a large, complex macromolecule, consisting largely of protein, and usually containing a *prosthetic* group, which often includes one or more metal atoms. In many enzymes, particularly those used in biosensors, the mode of action involves oxidation or reduction, which can be detected electrochemically. The detailed mode of action of enzymes can be found in any standard biochemistry text, so we will just remind ourselves here of the basic enzyme catalysis mechanism, which is as follows:

$$\mathrm{S} + \mathrm{E} \underset{k_{-1}}{\overset{k_1}{\rightleftharpoons}} \mathrm{ES} \xrightarrow{k_2} \mathrm{E} + \mathrm{P} \tag{3.2}$$

where S is the substrate, E is the enzyme, ES is the enzyme–substrate complex, and P is the product.

For example, if S is glucose, E is glucose oxidase (GOD), and P is gluconic acid, we have:

$$\text{Glucose} + \text{oxygen} + \text{GOD} \rightleftharpoons (\text{ES}) \longrightarrow \text{GOD} + \text{gluconic acid} + \mathrm{H_2O_2}$$

Let us now apply the steady-state approximation of the kinetic theory to the reaction system shown in equation (3.2). This approximation simply assumes that, during most of the time of the reaction, the concentration of the enzyme–substrate complex is steady, i.e. constant, so the rate of formation of the complex from its components is balanced by the rate of its breakdown back to enzyme and forward to its products. Thus, we have:

$$\text{rate of formation of complex} = k_1[\mathrm{S}][\mathrm{E}] - k_{-1}[\mathrm{ES}]$$

$$\text{rate of breakdown of complex} = k_2[\mathrm{ES}]$$

These rates are equal and opposite because of the steady-state approximation. Therefore, we can write:

$$k_1[\mathrm{S}][\mathrm{E}] - k_{-1}[\mathrm{ES}] - k_2[\mathrm{ES}] = 0$$

We describe the enzyme concentration in terms of the total $[E_0]$, rather than the unknown, [E], so that $[E_0] = [E] + [ES]$. Therefore:

$$k_1[S][E_0] - k_1[S][ES] - k_{-1}[ES] - k_2[ES] = 0$$

If we now solve this equation for [ES], we obtain the following:

$$[ES] = \frac{k_1[S][E_0]}{k_{-1} + k_2 + k_1[S]}$$

If we now put $K_M = (k_{-1} + k_2)/k_1$, where K_M is the *Michaelis constant*, we obtain:

$$[ES] = \frac{[E_0][S]}{K_M + [S]}$$

Then, the overall rate of the reaction (rate of formation of products) is given by the *Michaelis–Menton* equation, as follows:

$$v = \frac{d[P]}{dt} = \frac{-d[S]}{dt} = k_2[ES] = \frac{k_2[E_0][S]}{K_M + [S]}$$

When $[S] \gg K_M$, a maximum value of the rate constant, V_{max}, is reached, so that $V_{max} = k_2[E_0]$, and when $[S] = K_M$, $v = V_{max}/2$. This is illustrated in Figure 3.16, which shows a curve. It is experimentally more convenient to plot the data in a straight-line form, and inverting the Michaelis–Menton equation can achieve this, as follows:

$$1/v = \frac{K_M + [S]}{k_2[E_0][S]} = \frac{K_M}{k_2[E_0][S]} + \frac{1}{V_{max}}$$

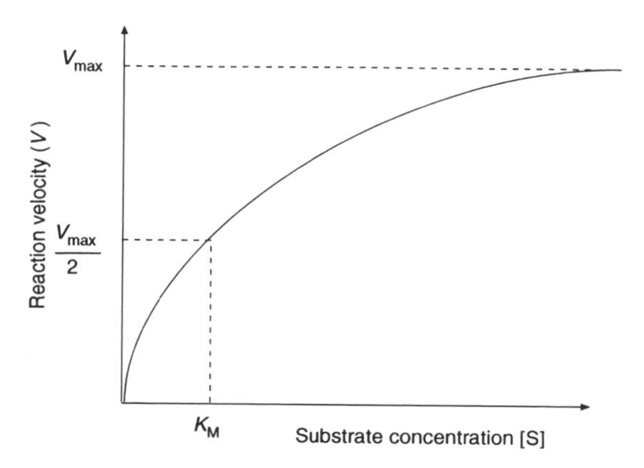

Figure 3.16 Dependence of reaction rate on substrate concentration for an enzyme-catalysed reaction at constant enzyme concentration. From Eggins, B. R., *Biosensors: An Introduction*, Copyright 1996. © John Wiley & Sons Limited. Reproduced with permission.

This is known as a *Lineweaver–Burk* plot, i.e. when $1/v$ is plotted against $1/[S]$, a straight line is obtained with a slope of K_M/V_{max} and an intercept of $1/V_{max}$. Hence, both K_M and V_{max} can be obtained. In practice, more steps may be involved and the effects of inhibitors would need to be considered.

The glucose biosensor and the urea biosensor are common examples of the use of enzymes as selective agents (see Chapter 5 for further details)

DQ 3.5

What are the advantages and disadvantages of using enzymes in sensors?

Answer

Advantages

(i) *They bind to the substrate.*
(ii) *They are highly selective.*
(iii) *They have catalytic activity, thus improving the sensitivity.*
(iv) *They are fairly fast-acting.*
(v) *They are the most commonly used biological components.*

Disadvantages

(i) *They are expensive. The cost of extracting, isolating and purifying enzymes is very high, and sometimes the cost of the source of the enzyme may also be high. However, a very wide range of enzymes is available commercially, usually with well-defined and assayed characteristics.*
(ii) *There is often a loss of activity when they are immobilized on a transducer.*
(iii) *They tend to lose activity, owing to deactivation, after a relatively short period of time.*

SAQ 3.7

Why does the response of an enzyme become non-linear at high analyte concentrations?

SAQ 3.8

Why is an enzyme not always totally specific for one reaction?

3.5.3 Tissue Materials

Plant and animal tissues may be used directly with minimal preparation. Generally, tissues contain a multiplicity of enzymes and thus may not be as selective as purified enzymes. However, the enzymes exist in their natural environment and so are less subject to degradation. Hence, the sensors derived from them are likely to have a longer lifetime. On the other hand, the response may be slower as there is more tissue material for the substrate to diffuse through. Such material may also dilute the effect of the enzymes. We shall look at a number of examples, comparing the performance criteria of biosensors made from tissue material with those made from pure enzymes.

Both micro-organisms and tissues are enzyme-containing materials, but the different environments surrounding the enzymes result in different advantages and disadvantages. Both materials are cheaper than isolated enzymes, and both show improved lifetimes when used in biosensors. The enzyme activity is stabilized in a more natural environment, and they may be more stable to inhibition by solutes, pH and temperature changes. Their major disadvantage is some loss of selectivity as they often contain a mixture of enzymes, although the latter may be of a related type. Thus, banana tissue, which was developed by Sidwell and Reichnitz (1985) and subsequently by Wang and Lin (1988) for the determination of dopamine, a catcholamine found in the brain and containing a complex of polyphenolases which catalyse the oxidation of polyphenolic compounds, has been found by ourselves (Eggins, 1994) to be equally effective for the determination of catechol itself and for flavanols, which are a type of catechol found as flavourings in beers and wines. This lowered selectivity can be turned to advantage in some sorts of analysis. The following DQ summarizes the advantages and disadvantages of plant materials when employed as selective elements in biosensors.

DQ 3.6

Discuss the advantages and disadvantages of using plant materials as selective elements.

Answer

Advantages

(i) The enzymes are maintained in their natural environments.
(ii) The enzyme activity is stabilized.
(iii) They sometimes work where purified enzymes may fail.
(iv) They are much less expensive than purified enzymes.

Disadvantage

There may be interfering processes, i.e. there can be some loss of selectivity.

Some examples are given below in Chapter 5, including biosensors used for arginine and flavanols.

3.5.4 Micro-Organisms

Micro-organisms play an important part in many biotechnological processes in industry, in fields such as brewing, pharmaceutical synthesis, food manufacture, waste-water treatment and energy production. Many biosensors based on micro-organisms immobilized on a transducer have been developed to assist with the monitoring of these processes, and various others. Micro-organisms can assimilate organic compounds, thus resulting in a change in respiration activity, and can also produce electroactive metabolites.

The advantages of using such materials in biosensors include the following:

(i) They are a cheaper source of enzymes than isolated enzymes.

(ii) They are less sensitive to inhibition by solutes, and are more tolerant of pH and temperature changes.

(iii) They have longer lifetimes.

However, they do have a number of disadvantages, namely:

(i) They sometimes have longer response times.

(ii) They have longer recovery times.

(iii) Like tissues, they often contain many enzymes and so may have less selectivity.

Some examples of applications are given below in Chapter 5, including biosensors used for assimilable sugars and glucose.

3.5.5 Mitochondria

These sub-cellular multi-enzyme particles can act as effective biocatalytic components. They can sometimes be useful in improving sensor response and selectivity when the entire tissue lacks the necessary properties. Table 4.2 below shows the relative performance characteristics of a mitochondrial biosensor for glutamine.

3.5.6 Antibodies

Antibodies are perhaps the most versatile biological selective agents. An antibody (Ab) can be developed against almost any substance (the antigen (Ag)) and thus makes a highly selective material. Organisms develop antibodies which are proteins that can bind with an invading antigen and thus remove it from harm, as follows (Figure 3.17):

$$\text{Ab} + \text{Ag} = \text{Ab–Ag}$$

The affinity constant, $K = [\text{Ag–Ab}]/[\text{Ag}][\text{Ab}]$, is usually about 10^6. For a fixed concentration of antibody, the ratio of the free to bound antigen, [Ag]/[Ag–Ab],

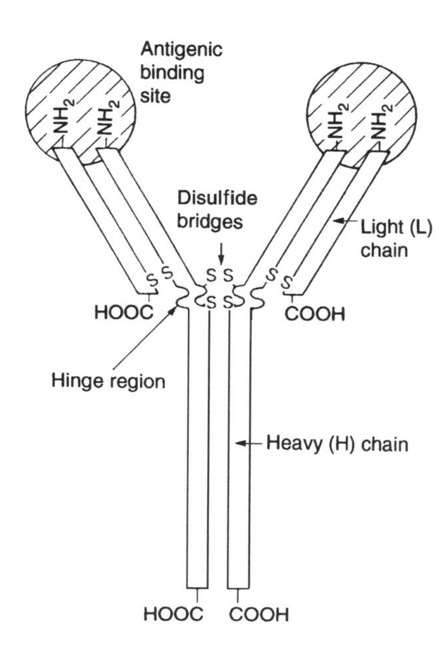

Figure 3.17 Scheme for a typical antibody composed of two heavy chains and two light chains. From Hall, E. A. H., *Biosensors*, Copyright 1990. © John Wiley & Sons Limited. Reproduced with permission.

at equilibrium is quantitatively related to the total amount of ligand, i.e. [Ag] + [Ag–Ab]. If a fixed amount of Ag is added to the assay, the unknown concentration of the original Ag can then be determined.

An unknown antibody can be determined by using labelled antibody material. Labelling may be carried out by using radioisotopes, enzymes, red (blood) cells, fluorescent probes, chemiluminescent probes or metal tags. In fact, the label can be present on either the Ab or the Ag. In biosensors, the emphasis has been on labelling with enzymes. Almost all transducers can be used with antibodies.

Such materials, when used as (selective) elements in biosensors, have the following advantages:

(i) They are very selective (in fact, they can be *too selective* between different strains).

(ii) They are ultra-sensitive.

(iii) They bind very powerfully.

However, a particular disadvantage is that there is *no catalytic effect*.

Antibodies have long been used in immunoassays. They bind even more powerfully and specifically to the corresponding antigen than enzymes do to their substrates. In fact, as indicated above, they can sometimes be too selective between different strains of the same material. They are ultra-sensitive, although they lack the catalytic activity of enzymes, and are often used in a labelled form (see above).

Some examples of the application of antibodies in biosensors are given below in Chapter 5, including sensors used for oestradiol and TNT.

SAQ 3.9

An antibody is always totally specific for one antigen. Why can this sometimes be a disadvantage?

3.5.7 Nucleic Acids

Nucleic acids operate in many ways like antibodies. The specific base-pairing between the strands of nucleic acids gives rise to the genetic code, which determines the replicating characteristics of all parts of living cells, and thus the inherited characteristics of individual members of a species. There is thus a genetically coded nucleic acid for each individual molecule created in, and by, a living cell, including proteins, and hence enzymes.

DNA probes can be used to detect genetic diseases, cancers and viral infections. They are used either in the 'short' synthetic form or in the 'long' form produced by cloning. As with antibodies, a DNA assay often involves the addition of labelled DNA to the system. The labelling may be radioactive, photometric, enzymic or electroactive, thus giving scope for a wide range of biosensor types.

With genetic and protein engineering, another use for DNA in developing sensors is possible. Regarding this, a number of approaches have been envisioned, including the following:

(i) *Improvement of enzyme yield.* Some useful enzymes are present in such small quantities that they are difficult to isolate. An example is glucose dehydrogenase (GDH), of potential use in glucose biosensors, which would have no requirement for oxygen (or other oxidant), unlike glucose oxidase. One source of GDH is the bacterium *Acinetobacter calcoaceticus*, although it is present only at low levels. These levels can be greatly enhanced by using cloning techniques, which effectively duplicate the gene encoding GDH, so producing up to 50 copies of the plasmid vector per cell and consequently large amounts of the enzyme. The details of this involve knowledge of genetic engineering techniques.

(ii) *Improvement of enzyme properties.* Changes in the enzyme itself can potentially be made, either directly by protein engineering or by modifying the replicating gene using genetic engineering. Modifications might (a) improve the turnover number of the enzyme, (b) change the pH dependence, (c) change the linear response to substrate concentration, (d) improve the stability during storage and operation, (e) reduce the susceptibility to interfering substances, (f) widen or narrow substrate specificity, and (g) change the cofactor requirement.

Considerable advances have been made in these areas, although there is not yet much regular application.

SAQ 3.10

Why are nucleic acids currently becoming of greater importance as bioselective agents?

3.5.8 Receptors

A receptor is a structure in a cell, which can trigger an amplified physiological response when it is bound to a particular ligand (an *agonist*). The response may be either (a) ion-channel opening, (b) production of second messenger systems, or (c) activation of enzymes. Such biological receptors tend to have an affinity for a range of structurally related compounds, rather than one specific analyte, which is often an attractive feature for use in biosensors. Frequently, they are used with labelled materials, originally with radioactive-labelled ligands, but now also with fluorescent- or enzyme-labelled ligands. Alternatively, the label could be on the receptor.

Sensors employing these systems can be broadly grouped into intact receptor-based biosensors and isolated receptor-based biosensors.

The examples given later in Chapter 5 include detectors for local anaesthetics, as well as those used for oestrogens and oestrogen mimics. In the former group, one of the most interesting experiments, resulting in a so-called prototype chemoreceptor, used excized crustacean olfactory structures for neuronal sensing of chemoreceptor ligands such as amino acids, sex hormones and pyridine-based materials. The binding of the ligand to the corresponding chemoreceptor triggers the firing of an action potential (AP) in the nerve, which can then be detected by a glass microelectrode outside of the cell.

A system using the giant axons of the crayfish was used to detect local anaesthetics, which can bind to the ion channel of the voltage-gated sodium channel receptor responsible for AP propagation, thus inhibiting the AP conduction. This system can also be applied to neuromodulator drugs or toxins, including antidepressants, narcotics, alcohols and venoms. The latter materials all affect the AP conduction. Unfortunately, such systems have a lifetime of only 4 to 8 h.

One of the most studied of the isolated receptors is nicotinic acetylcholine. This compound has been used with several transducers, including both ISFET and capacitance types. However, the most successful application has involved the use of a fibre-optic evanescent wave with fluorescent-labelled ligands (see Chapter 5 below). This has been used with fluorescein isothiocyanate (FITC)-labelled α-conotoxin, with this sensor having a lifetime of 30 days. There are many other interesting aspects of receptors to be investigated before successful biosensors based on these systems can be fully developed.

SAQ 3.11

How do receptors differ from antibodies in their mode of action?

3.6 Immobilization of Biological Components

3.6.1 Introduction

In order to make a viable biosensor, the biological component has to be properly attached to the transducer. This process is known as *immobilization*. There are five main methods of doing this, as follows:

(i) *Adsorption*. This is the simplest approach and involves minimal preparation. However, the bonding is weak and this method is only suitable for exploratory work over a short time-span.

(ii) *Microencapsulation*. This was the method used in the early biosensors. In this technique, the biomaterial is held in place behind a membrane, thus giving close contact between the biomaterial and the transducer. Such a method is adaptable, does not interfere with the reliability of the enzyme, and limits contamination and biodegradation. It is also stable towards changes in temperature, pH, ionic strength and chemical composition. However, such systems may be permeable to some materials, e.g. small molecules, including gases, and electrons.

(iii) *Entrapment*. Here, the biomaterial is mixed with a monomer solution, which is then polymerized to a gel, thus trapping the material. Unfortunately, this can cause barriers to the diffusion of substrate, thus slowing the reaction. It can also result in loss of bioactivity through pores in the gel – this effect can be counteracted however, by cross-linking (see below). The most commonly used gel is polyacrylamide, although starch, nylon and silastic gels have also been used. Conducting polymers, such as polypyrroles, are particularly useful with electrodes.

(iv) *Cross-linking*. In this method, the biomaterial is chemically bonded to solid supports or to another supporting material, such as a gel. Bifunctional reagents, such as glutaraldehyde, can be used in such techniques. Again,

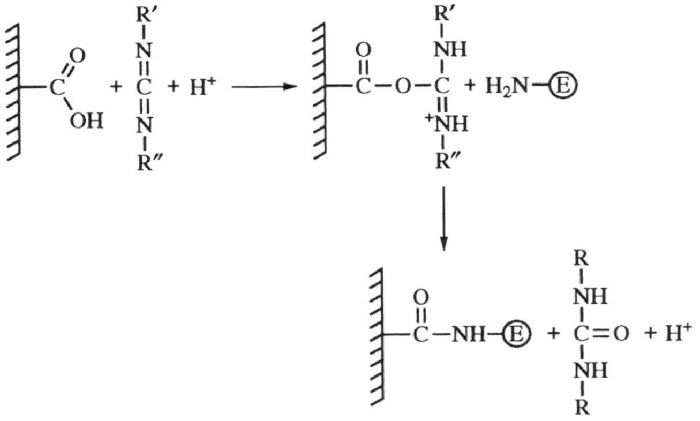

Figure 3.18 Covalent bonding of an enzyme to a transducer via a carbodiimide. From Eggins, B. R., *Biosensors: An Introduction*, Copyright 1996. © John Wiley & Sons Limited. Reproduced with permission.

there is some diffusion limitation and there can also be damage to the biomaterial. In addition, the mechanical strength of the system is poor. However, it can be a useful method for stabilizing adsorbed biomaterials.

(v) *Covalent bonding*. This approach involves a carefully designed bond between a functional group in the biomaterial and the support matrix. Nucleophilic groups in the amino acids of the biomaterials, which are not essential for the catalytic action of, say, an enzyme, are suitable in this respect. Many examples are known, but we will just illustrate the technique with one of these. In this case, a carboxyl group on the support is reacted with a carbodiimide. This then couples with an amine group on the biomaterial to form an amide bond between the support and the enzyme, as shown in Figure 3.18. The reaction must work under conditions of low temperature, low ionic strength, and neutral pH, and in this way the enzyme will not be lost from the biosensor device when using this technique.

Overall, the lifetime of the biosensor is greatly enhanced by proper immobilization. Typical lifetimes for the same biosensor, in which different methods of immobilization are used, are as follows:

Adsorption	1 day
Membrane entrapment	1 week
Physical entrapment	3–4 weeks
Covalent entrapment	4–14 months

SAQ 3.12

List the main methods for the immobilization of biological materials to transducers in order of their stabilities.

3.6.2 Adsorption

Many substances adsorb enzymes on their surfaces, e.g. alumina, charcoal, clay, cellulose, kaolin, silica gel, glass and collagen. No reagents are required, there is no clean-up step and there is less disruption to the enzymes.

Adsorption can roughly be divided into two forms, namely physical adsorption (*physisorption*) and chemical adsorption (*chemisorption*). Physisorption is usually weak and occurs via the formation of van der Waals bonds, occasionally including hydrogen bonds or charge-transfer forces. Chemisorption is much stronger and involves the formation of covalent bonds.

Several model equations are used to describe adsorption, but the most generally used is the *Langmuir adsorption isotherm*. This is derived from kinetic considerations and relates the fraction of the surface covered by the adsorbent (θ) with various kinetic parameters, as follows:

$$\text{rate of adsorption} = k_a p_a N(1 - \theta)$$

$$\text{rate of desorption} = k_d N\theta$$

At equilibrium the two rates are equal, and therefore:

$$\theta = Kp_a/(1 + Kp_a)$$

where p_a is the pressure of adsorbent, k_a the rate constant for adsorption, k_d the rate constant for desorption, and K is equal to k_a/k_d.

Adsorbed biomaterial is very susceptible to changes in pH, temperature, ionic strength and the substrate. However, this approach has proved to be satisfactory for short-term investigations.

SAQ 3.13

Why is adsorption of limited use as a method for immobilization?

3.6.3 Microencapsulation

In this method, an inert membrane is used to trap the biomaterial on to the transducer. This was the technique originally used to develop the first glucose biosensor on the oxygen electrode.

The advantages of this approach are as follows:

(i) There is close attachment between the biomaterial and the transducer.

(ii) It is very adaptable, and also very reliable.

(iii) The reliability of the biomaterial (enzyme) is maintained as follows:

  (a) a high degree of specificity is achieved;

  (b) there is good stability to changes in temperature, pH, ionic strength, E^0 and substrate concentration;

  (c) it can act as an inbuilt device to limit contamination and biodegradation;

  (d) if used with a patient, infection can be avoided.

(iv) There is always the option to bond the biological component to the sensor via molecules that conduct electrons, such as polypyrrole.

The main types of membranes used include cellulose acetate (dialysis membranes), which excludes proteins and slows the transportation of interfering species such as ascorbate, polycarbonate ('Nuclepore'), a synthetic material which is non-permselective, collagen, a natural protein, and polytetrafluoroethylene (PTFE) (often known under the tradename 'Teflon'), a synthetic polymer which is selectively permeable to gases such as oxygen. 'Nafion' and polyurethanes are among the other materials which are sometimes used for membranes.

3.6.4 Entrapment

In this approach, a polymeric gel is prepared in a solution containing the biomaterial. The enzyme is thus trapped within the gel matrix. The most commonly used polymer is polyacrylamide, which is prepared by the copolymerization of acrylamide with N,N'-methylenebisacrylamide. Polymerization can be effected by UV irradiation in the presence of vitamin B_1 as a photosensitizer. Other materials which have been used include starch gels, nylon, silastic gels and conducting polymers (such as polypyrrole).

The problems encountered with this method include the following:

(i) Large barriers are created, thus inhibiting the diffusion of the substrate, which slows down the reaction, and hence the response time of the sensor.

(ii) There is loss of enzyme activity through the pores in the gel, although this may be overcome by cross-linking, e.g., with glutaraldehyde.

3.6.5 Cross-Linking

This approach uses bifunctional agents to bind the biomaterial to solid supports, and has proven to be a useful method for stabilizing adsorbed enzymes. It does, however, have the following disadvantages:

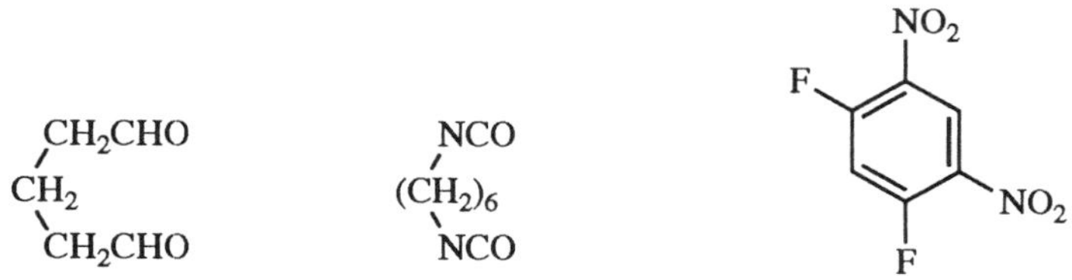

CH_2CHO / CH_2 \ CH_2CHO — **Glutaraldehyde**

NCO / $(CH_2)_6$ \ NCO — **Hexamethyl diisocyanate**

NO_2, F, NO_2, F — **1,5-Dinitro-2,4-difluorobenzene**

Figure 3.19 Some molecules used in cross-linking reactions. From Eggins, B. R., *Biosensors: An Introduction*, Copyright 1996. © John Wiley & Sons Limited. Reproduced with permission.

(i) Damage is caused to the enzyme.

(ii) Diffusion of the substrate is limited.

(iii) There is poor rigidity (mechanical strength).

Figure 3.19 shows the structures of the materials which are most commonly used for cross-linking reactions, i.e. glutaraldehyde (which will react with lysine amino acid residues in the enzyme), hexamethylene diisocyanate and 1,5-dinitro-2,4-difluorobenzene.

3.6.6 Covalent Bonding

Some functional groups, which are not essential for the catalytic activity of an enzyme, can be covalently bonded to the support matrix (transducer or membrane). This method uses nucleophilic groups for coupling, such as NH_2, CO_2H, OH, C_6H_4OH and SH, as well as imidazole.

Figure 3.20 shows several examples of the most common reactions used in this approach. Such reactions need to be performed under mild conditions, i.e. low temperatures, low ionic strengths and pH levels in the physiological range.

The particular advantage of this method is that the enzyme will not be released during use. In order to protect the active site, the reaction is often carried out in the presence of the substrate.

In practice, it is unusual for only one of the methods described above to be used at a time, as some of the following DQs and SAQs will illustrate.

DQ 3.7

Discuss the best combination of methods for use in a commercial sensor.

Answer

Robust stability is an essential for a commercial sensor, combined with reproducibility to match the performance required for the particular application. Cost is likely to be less important in the immobilization of a

(a) The cyanogen bromide technique

—OH
—OH
CNBr →
—O
C=NH
—O
H_2N—(E) →
O
||
—O—C—NH—(E)
—OH

(b) The carbodiimide method

O
//
—C
\
OH
+
R′
|
N
||
C
||
N
|
R″
+ H^+ →
O
||
—C—O—C
R′
|
NH
|
C
||
$^+$NH
|
R″
+ H_2N—(E)

↓

O
||
—C—NH—(E)
+
R
|
NH
|
C=O
|
NH
|
R
+ H^+

(c) Via acyl groups by treatment of hydrazides with nitrous acid

—C—CH_2—COOH
CH_3OH →
H^+
O
||
—C—CH_2—C—OCH_3

↓ H_2N—NH_2

O
||
—O—CH_2—C—N_3
$NaNO_2$ ←
H^+
O
||
—C—CH_2—C—NH—NH_2

↓ NH_2 (E)

O
||
—O—CH_2—C—NH
|
(E)

Figure 3.20 Some common reactions used for covalent bonding. From Hall, E. A. H., *Biosensors*, Copyright 1990. © John Wiley & Sons Limited. Reproduced with permission.

(d) Coupling using cyanuric chloride

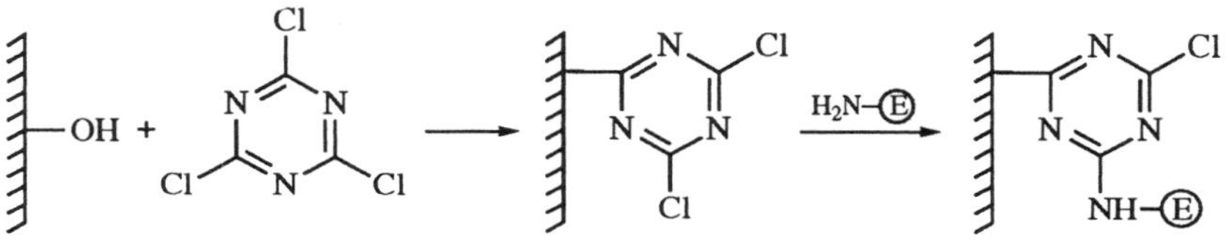

(e) Coupling through diazonium groups from aromatic amino groups

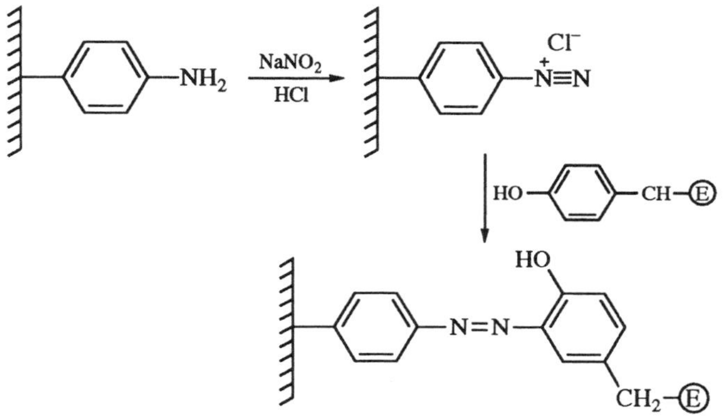

(f) Coupling via thiol groups

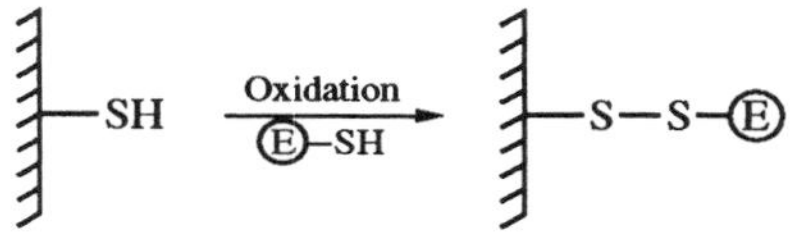

Figure 3.20 (Continued).

biological component than in other aspects, such as the performance of the transducer and the read-out. The covalent-bonding approach would be the most suitably robust method, followed by microencapsulation (depending on the biomaterial being used). However, if the sensor tip was to be disposable, the latter could be screen-printed and entrapment in a polymer matrix might then prove suitable.

SAQ 3.14

Which of the above methods are most suited to the immobilization of antibodies?

SAQ 3.15

Which of these methods are most suited for use with bacteria?

Summary

This chapter looks at the different elements which can sense analytes with some degree of selectivity. It discusses the various ways of sensing ions, and then molecules. Finally, methods of immobilizing the sensing element to the transducer are surveyed.

Ions are principally sensed by ion-selective electrodes, which are potentiometric devices. The uses of conductivity and electrochemical impedance spectroscopy are then considered. Modified electrodes and polymer-coated electrodes are discussed, and photometric methods of detecting ions are presented. Ion-selective electrodes are the most developed devices for the determination of particular ions. The selectivity limitations of the latter are considered, including the use of the Nicholskii–Eisenman equation.

Conductivity is a somewhat under-used method, which does not usually possess much specificity. Field-effect transistors must also normally be used with another selective element, e.g. an ISFET, CHEMFET or BIOFET. Modified electrodes are discussed, including those involving carbon paste, (modified) polymers, electronically conducting polymers, ion-exchange polymers and redox polymers.

Molecules can be detected by direct chemical methods, such as via the thermodynamic equilibrium in complex formation (this also applies to ions). They will also respond at different rates in chemical reactions, especially under the influence of a selective catalyst.

The main spectroscopic methods used in the identification of molecular species are outlined, including infrared, ultraviolet and nuclear magnetic resonance spectroscopies, as well as mass spectrometry. These techniques are not usually considered in the context of sensors, so for further details the interested reader should consult specialist texts.

When biological sensing elements are used, the sensor is known as a biosensor. Enzymes and antibodies have proved to be very successful sensing elements, while receptors and DNA are increasingly coming into use. The five main biological materials employed as selective agents in sensors, namely enzymes, antibodies, mitochondria, receptors and nucleic acids, are described and their modes of action are presented. The uses of enzyme-containing bacteria and organic tissue materials are also discussed. The advantages and disadvantages of each of these systems are given, and some examples of applications are outlined.

The five main methods for the immobilization of biological materials on to transducers are described. These are adsorption, microencapsulation, entrapment, cross-linking and covalent bonding. A range of methods of attaching the sensing element to the transducer has been developed, with some of these being used in combination. Adsorption is the simplest method, while the use of a membrane (microencapsulation) is very popular. Entrapment of the material in a matrix, which may be polymeric, is also being increasingly employed. For secure linking, chemical bonding, either between the individual molecules of the material (cross-linking) or direct to the transducer, may be used. Examples of applications of all of these approaches are given.

Further Reading

Arnold, M. A. and Reichnitz, G. A., 'Biosensors based on plant and animal tissues', in *Biosensors: Fundamentals and Applications*, Turner, A. P. F., Karube, I. and Wilson, G. S. (Eds), Oxford University Press, Oxford, UK, 1987, pp. 30–59.

Eggins, B. R., 'Recent advances in the fabrication and application of sensors', in *Proceedings of Conference on Analytical Advances in the Biosciences*, University of Ulster, UK, 23–24 June, 1994, p. 16.

Green, M. J., 'New approaches to electrochemical immunoassays', in *Biosensors: Fundamentals and Applications*, Turner, A. P. F., Karube, I. and Wilson, G. S. (Eds), Oxford University Press, Oxford, UK, 1987, pp. 60–70.

Karube, I., 'Micro-organism based sensors', in *Biosensors: Fundamentals and Applications*, Turner, A. P. F., Karube, I. and Wilson, G. S. (Eds), Oxford University Press, Oxford, UK, 1987, pp. 13–29.

Kuan, S. S. and Guilbault, G. G., 'Ion selective electrodes and biosensors based on ISEs', in *Biosensors: Fundamentals and Applications*, Turner, A. P. F., Karube, I. and Wilson, G. S. (Eds), Oxford University Press, Oxford, UK, 1987, pp. 135–152.

Leech, D., 'Affinity biosensors', *Chem. Soc. Rev.*, **23**, 205–213 (1994).

Sidwell, J. S. and Rechnitz, G. A., "'Bananatrode' – an electrochemical biosensor for dopamine', *Biotechnol. Lett.*, **7**, 419–422 (1985).

Vadgama, P. and Crump, P. W., 'Biosensors: recent trends. A review', *Analyst*, **117**, 1657–1670 (1992).

Wang, J. and Lin, M. S., 'Mixed plant tissue–carbon paste electrode', *Anal. Chem.*, **60**, 1545–1548 (1988).

Warner, P. G., 'Genetic engineering', in *Biosensors: Fundamentals and Applications*, Turner, A. P. F., Karube, I. and Wilson, G. S. (Eds), Oxford University Press, Oxford, UK, 1987, pp. 100–112.

Chapter 4

Performance Factors

Learning Objectives

- To appreciate that all analytical techniques have to be tested in order to determine their operating criteria.
- To appreciate that sensors must have a selectivity element that may be subject to interference effects.
- To understand the relative merits of different methods for minimizing interferences.
- To know the range, linear range and detection limits of particular sensors.
- To appreciate response times, recovery times and lifetimes.
- To distinguish between precision, accuracy and reproducibility.
- To be able to apply these principles to potentiometric and amperometric sensors.

4.1 Introduction

As a new technique is developed, one needs to establish fairly quickly the criteria by which its performance can be measured. These criteria have to be refined continuously as expectations are raised. This is especially true for a device containing biological materials. Showing that a method will work in a laboratory is a long way from delivering a commercial product that is reliable in the hands of (say) a medical laboratory technician or a nurse.

A number of different factors will be referred to in this chapter that apply to both biosensors and chemical sensors. Unfortunately, different workers express these factors in different ways, and worse still, some do not give much data at all about their performance in published papers.

Anyone developing a new sensor needs to have some idea of what performance requirements are likely to be needed for a particular application

4.2 Selectivity

This factor is the essence of sensors – it is their *raison d'etre*. It is rare to find a sensor which will respond to only one analyte, although some do exist. It is more usual to find a sensor that will respond mainly to one analyte, with a limited response to other similar analytes. Alternatively, the response may be to a group of analytes of similar chemical structure, such as carbonyl compounds.

SAQ 4.1

Distinguish between the terms 'specific' and 'selective'.

4.2.1 Ion-Selective Electrodes

Ion-selective electrodes respond to particular ions and nearly all are subject to interference from other similar ions. This interference can be measured and data are usually given in tables (see Table 4.1), or with the literature accompanying a commercial electrode. The extent of interference is expressed in the Nicolskii–Eisenman equation in terms of the electrode potential and a selectivity coefficient, $k_{i,j}$, as follows:

$$E = K + S\log(a_i + k_{i,j}a_j^{n/z}) \quad (4.1)$$

where a_i is the activity of the primary analyte of charge n and a_j the activity of the interfering analyte of charge z.

Table 4.1 Ranges and selectivity coefficients of some ion-selective electrodes

Ion	Range (M)	Selectivity coefficient, k_{ij}
Fluoride	10^0-10^{-6}	$OH^- = 0.1$
Chloride	10^0-10^{-4}	Br^- and I^- must be absent
Bromide	$10^0-5 \times 10^{-6}$	$Cl^- = 2.5 \times 10^{-3}$; $OH^- = 3 \times 10^{-5}$; I^- and S^{2-} must be absent
Iodide	10^0-10^{-6}	$CN^- = 1.0$; S^{2-} must be absent
Nitrate	10^0-10^{-6}	$ClO_4^- = 100$; $ClO_3^- = 100$
Sulfide	10^0-10^{-7}	Hg and Ag must be absent
Sodium	10^2-10^{-6}	$K^+ = 3 \times 10^{-2}$; $NH_4^+ = 2 \times 10^{-2}$
Potassium	10^0-10^{-5}	$Na^+ = 2.6 \times 10^{-3}$; $NH_4^+ = 0.3$; $Li^+ = 2.1 \times 10^{-3}$
Calcium	10^0-10^{-5}	$Na^+ = 3.3 \times 10^{-3}$; $Mg^{2+} = 0.015$; $Zn^{2+} = 1.2$; $K^+ = 2.2 \times 10^{-5}$
Silver	10^0-10^{-7}	Hg and S^{2-} must be absent

The value of $k_{i,j}$ can be determined by a 'two-point, mixed solution method'. The cell potential is measured in (say) a 0.001 M solution of the primary analyte alone (E_1) and then in a solution containing 0.001 M of the primary analyte together with (say) 0.1 M or 0.01 M of the potentially interfering ion (E_2). The selectivity coefficient is then calculated as follows:

$$E_1 = K + S \log a_i \tag{4.2}$$

$$E_2 = K + S \log (a_i + k_{i,j} a_j^{n/z}) \tag{4.3}$$

If we now subtract equation (4.2) from equation (4.3) and take the anti-log, we obtain the following expression:

$$k_{i,j} = \frac{a_i 10^{(E_2 - E_1)/S} - a_i'}{a_j^{n/z}} \tag{4.4}$$

where S is the slope of the calibration graph for the primary analyte (ideally, $59/n$ mV) and a_i' is the corrected activity of the analyte in the mixed solution.

DQ 4.1

For a calcium ISE, the calibration slope, S, was +29.6 mV/decade in a 0.001 M solution. In a 0.001 M calcium chloride solution, the cell potential was −20.1 mV, while the potential in a mixed solution containing 0.001 M calcium chloride and 0.1 M sodium chloride was −19.8 mV.

Calculate the selectivity coefficient for calcium ions in the presence of sodium.

Answer

The activities are obtained for the components in the mixed solution by using the activity coefficients for calcium and sodium, which are 0.40 and 0.77, respectively. Hence, the activities in the mixed solution are 4.0×10^{-4} M and 7.7×10^{-2} M, respectively.
Thus:

$$\begin{aligned} k_{i,j} &= [4 \times 10^{-4} \times 10^{(-19.8+20.1)/29.6} - 3.977 \times 10^{-4}]/(0.077^{2/1}) \\ &= [4 \times 10^{-4} \times 1.0236 - 3.977 \times 10^{-4}]/(0.077^{2/1}) \\ &= 2.0 \times 10^{-3} \end{aligned}$$

which shows sodium to be weakly interfering.

Looking at Table 4.1 we see that $k_{i,j}$ for hydroxide in the presence of a fluoride ion electrode is 0.1. This tells us (approximately) that a solution containing 10 times as much hydroxide as fluoride would give a response double that of the value for fluoride alone, so that hydroxide is a major interferent in this case.

However, the latter can easily be removed by buffering the test solution at about pH 5 when the activity of hydroxide would be 10^{-9}, which is 1/1000 of the normal detection limit of the electrode.

With all ion-selective electrodes it is important to use an ionic-strength adjustment buffer (ISAB) so that the ionic strengths of all of the solutions being measured are the same. In the case of some ions, extra precautions need to be taken, such as adjusting the pH (as with fluoride). Another problem with fluoride is that it complexes strongly with any aluminium or ferric ions in the solution. Therefore an, additional buffering agent is used – citrate, which complexes more strongly with aluminium and ferric ions than fluoride will do. Thus, for fluoride, the ISAB should consist of 0.1 M NaCl (to adjust the ionic strength) and 0.01 M sodium citrate (to remove Al and Fe), adjusted to pH 5.5 (to remove hydroxide).

4.2.2 Enzymes

Enzymes have been the most generally used selective agents in biosensors. Their selectivities depend very much on the actual enzymes being employed and their activities towards different analytes. They can be very selective in particular situations. Generally, glucose oxidase is highly selective for glucose in the presence of other sugars in a blood solution. However, the more nebulous polyphenoloxidases will catalyse the oxidation of a range of phenolic compounds. The common commercially available source, i.e. tyrosinase (obtained from mushrooms), will catalyse the oxidation of a range of phenolic compounds to different extents. However, the polyphenoloxidase obtained from bananas is reported not to catalyse tyrosine or phenol.

The broader-spectrum enzymes may be more useful in particular situations, for example, to give a general indication of the level of polyphenolic compounds in beers, which may cause haze.

4.2.3 Antibodies

Antibodies are the most selective biological components and can be developed, by monoclonal techniques, against any agonist. They are usually so selective that they will distinguish between *different isomers of the same compound*, let alone compounds of similar structure. However, this may be a disadvantage at times. Nevertheless, it is ideal for the highly selective determination of a particular component, such as antibiotic residues in cattle.

4.2.4 Receptors

Receptors are a relatively novel type of biosensor component. Such materials respond to an analyte that causes a particular effect on the receptor. An example of this is the oestrogen receptor, which will respond to any substance that will bind to it. Some of these are strongly oestrogenic, such as 17-β-oestradiol, while others are weaker, such as the oestrogenic mimics, alkylpolyphenol ethoxylates. Thus, a sensor based on such an acceptor will not respond equally to all oestrogenic compounds, but will very usefully give an indication of the extent of the oestrogenic activity of a given concentration of a particular compound. If one requires specific quantitative information about a particular compound, then the immunoassay method would be preferred.

4.2.5 Others

Greater selectivity can, of course, be obtained by spectroscopic methods and by combinations of chromatographic methods with other sensing detectors. Gas chromatography–mass spectrometry (GC–MS) and liquid chromatrography (LC)–MS are very powerful, although expensive methods. There is much scope for the use of chromatography, particularly high performance liquid chromatrography (HPLC) combined with a selective sensor as the detector. Thus, the phenolic components in beers can be distinguished by using this technique. Furthermore, the use of a polyphenoloxidase biosensor as the detector could enhance this method.

SAQ 4.2

What is the maximum tolerated concentration of hydroxide ions for a 10% error when using a fluoride ion ISE to measure fluoride ions in the range 10^{-3} to 10^{-6} M ($k_{ij\ OH^-} = 0.1$)? How can one guarantee the absence of interference from hydroxide ions?

4.3 Sensitivity

4.3.1 Range, Linear Range and Detection Limits

With any analytical technique, it is important to know what concentration range is covered and from the calibration point of view, over what section of this range the response is linear. At the lower level is the *detection limit*. This has a precise definition according to the IUPAC convention (see Figure 4.1). It is the concentration of analyte at which the extrapolated linear portion of the calibration graph intersects the baseline – a horizontal line corresponding to zero change in response for several decades of concentration change.

To be useful, the detection limit needs to be better than 10^{-5} M (0.01 mM). The importance of the range can be illustrated from the example of the determination of glucose in blood. This needs to be at least 0.2–20 mM (preferably, 1×10^{-4} to 5×10^{-2} M) to cover the likely blood glucose levels found in normal and diabetic persons.

Linear ranges are generally much larger for potentiometric sensors. Thus, the H^+ (pH)-selective electrode has a practical range from pH 0 to pH 12, i.e. covering twelve powers of ten of the hydrogen ion concentration. Other ion-selective

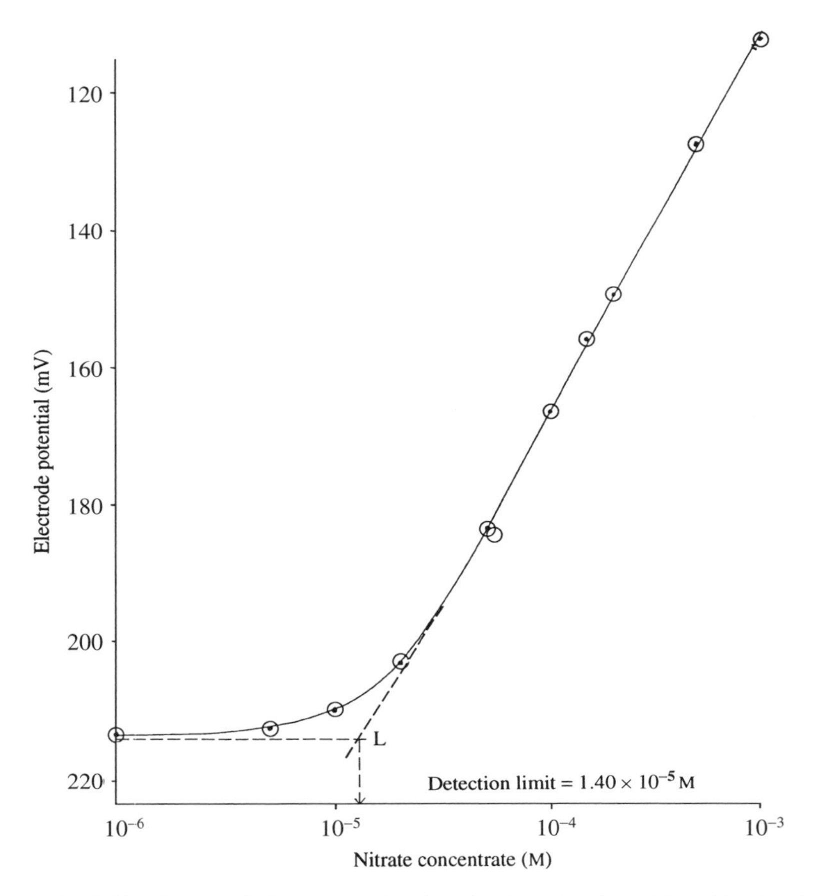

Figure 4.1 Calibration graph for nitrate, showing the detection limit. From Eggins, B. R., *Biosensors: An Introduction*, Copyright 1996. © John Wiley & Sons Limited. Reproduced with permission.

electrodes cover quite considerable (although much lower) ranges, e.g. many of them cover six or seven powers of ten (see Table 4.1). Amperometric sensors and biosensors generally do not have ranges of much more than two or three powers of ten (see Tables 4.2–4.4 below).

DQ 4.2

An amperometric sensor for copper was used to measure the copper levels in a series of waste-water samples, with the following data being obtained: 46, 50, 53, 47, 55, 52, 48, 46, 53 and 54 ppm.

Calculate (a) the mean value, (b) the standard deviation, and (c) the spread of values.

Answer

Sample	*Concentration (ppm)*	$(x_i - x)$	$(x_i - x)^2$
1	*46*	*4.4*	*19.36*
2	*50*	*0.4*	*0.16*
3	*53*	*2.6*	*6.76*
4	*47*	*3.4*	*11.56*
5	*55*	*4.6*	*21.16*
6	*52*	*1.6*	*2.56*
7	*48*	*2.4*	*5.76*
8	*46*	*4.4*	*19.36*
9	*53*	*2.6*	*6.76*
10	*54*	*3.6*	*12.96*
Mean	50.4		$\Sigma = 106.4$
		$106.4/9 = 11.8222$	
		$\sqrt{} = 3.438 =$ ***Standard deviation***	
		$\% = 3.348/(50.4 \times 100) = 6.8\%$	

(a) Mean $= 1/n\Sigma n_i = 1/10(46 + 50 + 53 + 47 + 55 + 52 + 48 + 46 + 53 + 54) = 50.4$

(b) Standard deviation $= \sqrt{[\Sigma(x_i - x)^2/(n - 1)]} = 3.44 (= \pm 6.8\%)$

(c) Spread $= 46 - 55$

> **Note:** *Scientific calculators can usually calculate these amounts directly, while some can also fit a set of data to the best straight line and calculate the slope, intercept and correlation coefficient.*

4.4 Time Factors

While it is desirable for a sensor to have a rapid response time and to recover rapidly, ready for the next reading, this is not always the case in practice. Most sensors, whether chemical or biological, have a limited lifetime, so it is important to understand when a sensor should be discarded.

4.4.1 Response Times

Many analytical devices require some 'settling-down' time, i.e. time to allow the system to come to equilibrium, the response time. This is true to some extent with many chemical methods – for example, recent extensive studies with nitrate electrodes showed that the most reproducible results were obtained after stirring the solution in contact with the electrode for ca. 30 s. With biosensors, such measurement times may be longer than this. Obviously, if the time becomes too long it can materially affect the usefulness of the method for repetitive routine analyses. However, with sensors of a chemical or biochemical nature this response time is greatly offset by the simplicity of the measurement and the minimal sample preparation time. For biosensors, the response times can vary from a few seconds to a few minutes. Up to 5 min may be acceptable, but if the time exceeds 10 min this may be too long.

4.4.2 Recovery Times

Very much related to the response time is the *recovery time* – the time that elapses before a sensor is ready to be used for another sample measurement. The resulting response time may be immediate or it may be that after one measurement the sensor system has to rest to resume its base equilibrium before it can be used with the next sample. In many publications, these times are combined and the result is given as the number of samples that can be analysed per hour, which obviously is the main practical point at the end.

4.4.3 Lifetimes

The *lifetime* of a sensor can be regarded in several different ways. Even the most robust pH electrodes tend to deteriorate after months of use. First, we can define the response during continuous use, i.e. the sensor is in constant contact with an analyte solution and successive readings are made over a period of, e.g. hours. This lifetime in use can be defined as the time after which the response has declined by a given percentage (say 5%).

The second lifetime represents the time over which the assembled sensor is stored, perhaps in a buffer solution or an ISAB. This lifetime will depend on the care of the user in following the manufacturer's instructions.

The third lifetime is the period when the sensor is stored dry in its packing, for an ion-selective electrode, or when a biological material is stored separately, perhaps refrigerated.

All organic material deteriorates with time, especially when taken out of its natural environment. This means that one of the main drawbacks of biosensors is that the biological material usually has a fairly limited lifetime before it needs replacing. All developments of new biosensors include studies to show how the response of the biosensor to a standard sample changes with time over hours, days and even months. Generally pure enzymes have the lowest stability, whereas tissue preparations have the highest.

A number of techniques are currently being investigated to improve this aspect of biosensor development. Gibson and others have succeeded in stabilizing a range of enzymes by using a mixture of a polyelectrolyte (diethylaminoethyl

(DEAE)-dextrin) and a sugar alcohol (lacticol) as soluble additives for enzyme solutions. This treatment enhanced the retention of enzyme activity in solution, during desiccation and during thermal stress. The principal enzymes studied were alcohol dehydrogenase and horseradish peroxidase. Enhanced performances were also obtained with twelve other enzymes.

Further studies have been made with biosensors constructed for alcohol determination, by using alcohol oxidase, one with membrane immobilization and amperometric oxidation of hydrogen peroxide and the other with a mediated coupled reaction with horseradish peroxidase and *N*-methylphenothiazine–tetracyanoquinodimethane (NMP–TCNQ) on a graphite electrode. In both cases, addition of the stabilizers promoted a considerable increase in storage stability of the enzyme component, as indicated by an increase in the shelf life when stored in the dried form at 37°C. Similarly, an L-glutamate biosensor made from NMP–TCNQ-modified graphite electrodes and L-glutamate oxidase also showed an increase in shelf life when stored desiccated in the presence of stabilizers.

Other workers, who attempted to use just the DEAE-dextran polyelectrolyte with glucose oxidase biosensors, with dimethylferrocene in carbon paste electrodes, were less successful. In fact, lyophilized glucose oxidase alone is extremely stable, i.e. it is stable for 2 years at 0°C and for 8 years at −15°C. In solution, it is most stable at pH 5, while below pH 2 and above pH 8 the catalytic activity is rapidly lost.

SAQ 4.3

What factors affect the lifetime of an enzyme biosensor?

4.5 Precision, Accuracy and Repeatability

For any analytical instrument, including sensors, the analytical value must have sufficient precision for the required purpose, i.e. the random errors must be below a certain level, so that repetitive measurements are reproducible within a certain range. The sensor must also be capable of measurements of values with an accuracy close to the expected value. This means that the systematic errors must be below certain limits. This can be a particular problem where biological selective elements are used, as one sample can differ from another, thus giving systematic errors. Sufficient controls and standards must be used to enable repeatable results of sufficient accuracy to be obtained over an extended period. This is fully discussed in many standard textbooks, e.g. the text by Miller and Miller (see the Bibliography).

No analytical result has any real significance unless one can estimate the probable error of the device, i.e. the sensor in this case. This applies even more so with biosensors. Several readings can be made for each determination, or much better, several replicate determinations can be made. Then, a standard deviation needs to be calculated – this can usually be carried out by using a scientific pocket calculator. If a calibration curve is to be plotted, the data may be fitted by a linear least-squares fit to give the best (i.e. calculated) straight line, with its slope, intercept, standard deviation and correlation coefficient (i.e. 'goodness of fit'). The latter can be done with a simple computer program, which is now often found on most standard scientific calculators, or via a spreadsheet, such as Microsoft 'Excel'. More ambitious, but very important, is to correlate one set of data with another. Again, standard statistical methods are available to determine the relative standard deviation. With biosensors, the expected reproducibility between replicate determinations should be at least $\pm(5 - 10)\%$.

Once a method is established, one way of avoiding the necessity for plotting a full calibration curve every time is to use the *standard addition method*, or better still, the *multiple standard addition method*. Here, the sample to be analysed is measured, and then a sample of a known standard with a concentration of about twice that expected for the unknown is added and a second measurement is made. This approach assumes that the sensor response is known to vary linearly with sample concentration.

Let the response r, for a concentration C be:

$$r = kC$$

Therefore, for the unknown, U, we have:

$$r_{\rm U} = kC_{\rm U} \tag{4.5}$$

Similarly, for the sample with the added standard S:

$$r_{\rm (U+S)} = k(C_{\rm U} + C_{\rm S}) \tag{4.6}$$

Dividing equation (4.6) by equation (4.5) gives the following:

$$r_{\rm (U+S)}/r_{\rm U} = (C_{\rm U} + C_{\rm S})/C_{\rm U} = 1 + C_{\rm S}/C_{\rm U}$$

and by rearranging, we obtain:

$$C_{\rm U} = r_{\rm U}C_{\rm S}/(r_{\rm (U+S)} - r_{\rm U})$$

With the multiple standard addition method, several aliquots of the standard are added to the system. A graph is then plotted of response against the amount of each addition, and the straight line obtained is then extrapolated back to zero response. The negative intercept on the concentration axis represents the unknown concentration (see Figure 4.2).

If the relationship between the analyte concentration and the sensor response is logarithmic (as with potentiometric-based sensors), it is best to take the anti-log and then produce a Gran plot:

$$E = K + S\log(C_{\rm U} + C_{\rm S})$$

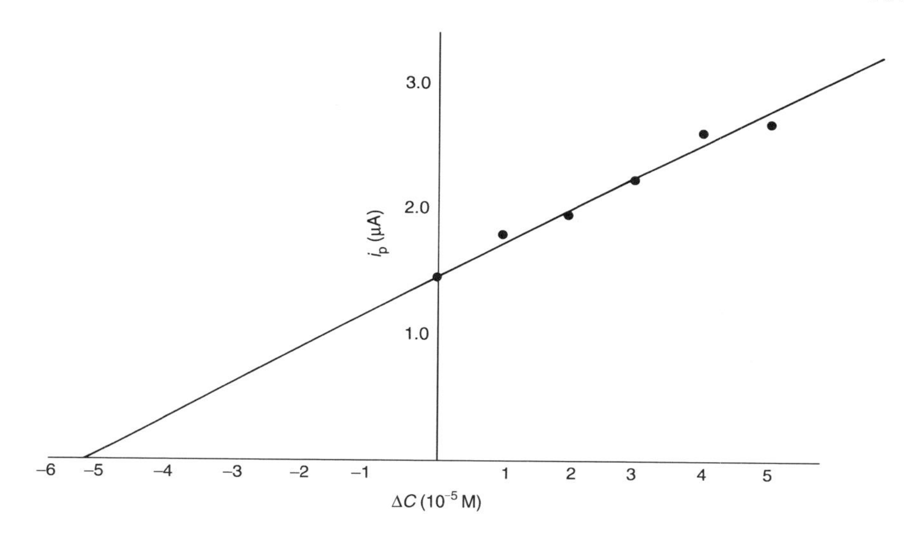

Figure 4.2 Multiple standard addition method, as applied to the determination of catechol in beer. From Eggins, B. R., *Biosensors: An Introduction*, Copyright 1996. © John Wiley & Sons Limited. Reproduced with permission.

and thus:

$$10^{E/S} = 10^{K/S}(C_U + C_S).$$

In this case, one plots $10^{E/S}$ versus C_S. The negative intercept at $10^{E/S} = 0$ gives C_U in a similar way to the multiple standard addition plot.

SAQ 4.4

Distinguish between 'accuracy', 'precision' and 'reproducibility'.

SAQ 4.5

How do potentiometric and amperometric sensors compare with regard to their linear ranges and precision?

4.6 Different Biomaterials

One of the best-studied systems is a biosensor for glutamine based on glutaminase, which is found in pork liver. This enzyme catalyses the deamination of glutamine to glutamate. The ammonia is detected potentiometrically with an

Table 4.2 Response characteristics of glutamine biosensors. Reprinted from **Biosensors: Fundamentals and Applications** edited by A. P. F. Turner, I. Karube and G. S. Wilson (1987), by permission of Oxford University Press

Parameter	Enzyme	Mitochondria	Bacteria	Tissue
Slope (mV per decade)	33–41	53	49	50
Detection limit (M)	6.0×10^{-5}	2.2×10^{-5}	5.6×10^{-5}	2.0×10^{-5}
Linear range (mM)	0.15–3.3	0.11–5.5	0.1–10	0.064–5.2
Response time (min)	4–5	6–7	5	5–7
Lifetime (days)	1	10	20	30

ammonia-sensitive electrode. This enzyme is also found in mitochondria and in bacteria. A comparison was made of the various response features for a biosensor made with (i) the isolated enzyme, (ii) a mitochondrial preparation, (iii) a bacterial preparation, and (iv) liver tissue. Some data for the factors discussed are given in Table 4.2. The advantages of the liver tissue are shown in almost all areas. The slope, i.e. 50 mV per decade, is closer to the optimum Nernstian value of 59 mV per decade than any of the others, while the detection limit is marginally better at 0.02 mM. The linear range, 0.064–5.2 mM, is wider and the lifetime of the sensor is dramatically longer – 30 days compared with 1 day for an enzyme sensor. However, the response time is marginally poorer, i.e. 5–7 min, compared to 4–5 min for the enzyme sensor.

4.7 Different Transducers

Performance is also affected by the amount of enzyme used. This is, of course, a general effect but, as can be seen in Table 4.3, it is particularly noticable in the cases of urea, glucose and penicillin. This table also shows some variation with the method of immobilization. e.g. physical entrapment in polyacrylamide gel (P), covalent bonding with glutaraldehyde and albumin to poly(acrylic acid) (or acrylamide), followed by physical entrapment (C), or dissolution (D).

Some examples of the above are given in the following sections.

4.7.1 Urea Biosensors

The most extensive comparisons have been carried out for the urea biosensors, for which data have been obtained for all of the usual modes of operation, namely cationic (NH_4^+), pH, and gas (NH_3 and (CO_2) (see Table 4.3).

The pH sensor needs the largest amount of enzymes and the response is linear only below 5×10^{-3} M. It also has the longest response time. The ammonia gas electrode is the best overall, being stable for 4 months, in addition to having the widest response range, using very little enzyme and having a fairly short response time. The cationic electrode has the best response time, with an excellent response range, although it is stable for only 3 weeks.

Table 4.3 Comparison of the performance characteristics of some potentiometric biosensors. Reprinted from **Biosensors: Fundamentals and Applications** edited by A. P. F. Turner, I. Karube and G. S. Wilson (1987), by permission of Oxford University Press

Type	Enzyme[a]	Sensor[b]	Stability	Response time (min)	Range (M)
Urea	Urease (25 U)	Cation (P)	3 weeks	0.5–1	10^{-1}–5×10^{-5}
Urea	Urease (75 U)	Cation (P)	4 months	1–2	10^{-2}–10^{-4}
Urea	Urease (100 U)	pH (P)	3 weeks	5–10	5×10^{-2}–5×10^{-5}
Urea	Urease (10 U)	Gas (NH_3)(C)	4 months	2–4	5×10^{-2}–5×10^{-5}
Urea	Urease (25 U)	Gas (CO_2)(P)	3 weeks	1–2	10^{-2}–10^{-4}
Glucose	GOD (100 U)	pH (D)	1 week	5–10	10^{-1}–10^{-3}
Glucose	GOD (10 U)	Iodide (C)	> 1 month	2–8	10^{-3}–10^{-4}
L-Amino acids					
General	L-AA oxidase	Cation (P)	2 weeks	1–2	10^{-2}–10^{-4}
		Iodide (C)	> 1 month	1–3	10^{-3}–10^{-4}
L-Tyrosine	L-Tyrosine carboxylase	Gas (CO_2) (P)	3 weeks	1–2	10^{-1}–10^{-4}
L-Glutamine	Glutaminase	Cation (D)	2 days	1	10^{-1}–10^{-4}
L-Glutamic acid	Glutamate dehydrogenase	Cation (D)	2 days	1	10^{-1}–10^{-4}
L-Aspargine	Asparginase	Cation (P)	1 month	1	10^{-2}–5×10^{-5}
D-Amino acids					
General	D-AA oxidase	Cation (P)	1 month	1	10^{-2}–5×10^{-5}
Penicillin	Penicillinase				
	(400 U)	pH (P)	1–2 weeks	0.5–1	10^{-2}–10^{-4}
	(1000 U)	pH (D)	3 weeks	2	10^{-2}–10^{-4}
Amygdalin	β-Glucooxidase	Cyanide (P)	3 days	10–20	10^{-2}–10^{-5}
Nitrate	Nitrate reductase	Ammonium (D)	1 day	2–3	10^{-2}–10^{-4}

[a] U, units of enzyme (urease) activity.

[b] P, physical entrapment in polyacrylamide gel; C, covalent bonding with glutaraldehyde and albumin to poly(acrylic acid) (or acrylamide), followed by physical entrapment; D, dissolution.

A urea biosensor based on an ENFET, with urease in the sensing gate, used changes in pH to measure the urea. This system was used in a differential mode so that the initial rate of voltage change was proportional to the logarithm of the urea concentration over the range 1.3–16.7 mM urea. If the sensor was stored at 4°C between measurements, it continued to work satisfactorily for at least 2 weeks. The response time was 2 min. Selectivity was tested against glucose, creatinine and albumin, and no interfering responses were observed. Another urea biosensor was based on an lr–Pd MOS (metal-oxide semiconductor) device, which at pH > 8 is fairly selective for ammonia. The urease (40 U) was contained in a flow column (40 × 2 mm id 'Eupergit'). A linear response was obtained up to 40 μM (±2%), with a detection limit of 0.2 M. The response rate permitted 20 assays per hour, while the urease column gave unchanged operation for one month.

4.7.2 Amino Acid Biosensors

Table 4.3 also shows some applications of potentiometric biosensors to the analysis of various amino acids. These mostly involve ammonium cation electrodes. There are some surprising differences between the general sensor for the D- and L-forms, where both use the appropriate amino acid oxidases and the same method of immobilization (polyacrylamide gel entrapment). The stabilities are given as 1 month for the D-forms, but only 2 weeks for the L-forms. The amount of enzyme needed is 50 U for the D-forms, but only 10 U for the L-forms. The concentration ranges are very similar, as are the response times (2 min). These biosensors will respond to D- and L-leucine, D- and L-methionine, and D- and L-phenylalanine, and also work with L-cysteine, L-tyrosine and L-tryptophan, and with D-alanine, D-valine, D-norleucine and D-isoleucine.

4.7.3 Glucose Biosensors

Potentiometric glucose electrodes are not used very much but are interesting systems for comparison with the more usual amperometric types. The pH mode has the best range but uses a considerable amount of enzyme and is stable for only one week. The iodide electrode allows a lower detection range, has a longer lifetime and uses fewer enzymes. For glucose determination in blood, a working range of (2×10^{-3})–(5×10^{-2}) M is desirable. There have been so many variations of the glucose electrode that an overall comparison is difficult. However, Table 4.4 shows the performance data obtained for some commercially produced glucose biosensors, including the original 'Yellow Spring Instruments Model 23A'. The lifetime data should be viewed with caution as the term 'lifetime' may signify different things with different biosensors, as discussed above in Section 4.4.3. The earliest glucose biosensors monitored oxygen, but for blood analysis the blood oxygen interferes, and so the biosensors to be used for clinical work were based on hydrogen peroxide measurements. Some mediated sensors are included in Table 4.4, in particular, the 'Medisense ExacTech' sensor based on ferrocene.

Table 4.4 Comparison of the performance factors of some glucose biosensors

Type	System	Range	Response time (min)	Lifetime
Glucorecorder (Analytical Instrument, Japan)	GOD–O_2	0–5 mM (±2%)	0.5	—
Radelkkis (Hungary)	OP-G1–7113-S	1.7–2.0 mM (±5%)	1.5	8 months
Yellow Springs Instruments, Model 23A (USA)	GOD–H_2O_2	1.0–45.0 mM (±2%)	1.5	300 samples
Glukometer, GKM01 (ZWG Academy of Sciences, Germany)	GOD–H_2O_2	0.5–50 mM (±1.5%)	0.7–1	1000 samples
Glucose Analyser 5410 (Hofmann-LaRoche, Switzerland)	GOD–$[Fe(CN)_6]^{3-}$	2.5–27.5 mM (±1.5%)	1	8 weeks
ExacTech (Medisense, UK)	GOD–ferrocene	1–30 mM (±1%)	0.5	> 1 year
	GOD–TTF–TCNQ	0.5–20 mM	—	100 days
	GDH–PQQ	1–70 mM	< 0.3	8 h
	Con A–fluorescent dextran (optode)	2.0–25 mM (±0.5%)	—	15 days
	Hexakinase–bacterial luciferase–ATP	2–100 pmol	—	—
	Hexokinase–O_2–thermistor	0.5–25 mM (±0.6%)	1.5	—

The advantages of the tetrathiafulvalene (TTF)–TCNQ-mediated and glucose dehydrogenase–pyrroloquinolinequinone (GDH–PQQ) sensors are not immediately apparent from the data. The glucose dehydrogenase–PQQ sensors produce a much larger current than the glucose oxidase (GOD)-based sensors and are not affected by ambient oxygen. The enzyme is effectively 'wired' directly to the electrode.

4.7.4 Uric Acid

A number of comparative studies have been made of biosensors for uric acid, a knowledge of which is important in haematology disorders. The normal operating range is 140–420 mol l^{-1}. Uric acid is oxidized in the presence of uricase by oxygen, according to the following:

$$\text{Uric acid} + O_2 \longrightarrow \text{allantoin} + CO_2 + H_2O_2$$

As with glucose, several modes of operation are possible, as follows:

(i) Direct measurement of oxygen – linear up to 0.5 mM (±5%) and operational for up to 100 days.

(ii) Measurement of hydrogen peroxide – linear up to 3.0 mM (±2%), and operational for 17 days (1000 samples).

(iii) Use of a mediator, $[Fe(CN)_6]_3$, via horseradish peroxidase – linear up to 0.035 mM, and operational for 40 days.

DQ 4.3

Discuss the performance factors of glucose biosensors for use in human blood analysis with regard to the following:

(a) Linear range;
(b) Response time;
(c) Precision.

Answer

(a) The linear range must cover not only the normal range of glucose concentrations in blood but also the possible abnormal ones. A range of 1×10^{-4} *M to* 5×10^{-2} *M is required.*
(b) This is a very frequently used assay, often operated by the patient, so the response time should be better than 30 s.
(c) The precision (reproducibility) should be much better than the usual ±5 %, requiring a value of about ±1 % for such an important medical measurement.

4.8 Some Factors Affecting the Performance of Sensors

4.8.1 Amount of Enzyme

Enzymes are catalysts, i.e. not consumed by the reaction, and the precise amount (or concentration) is not crucial for the operation of a biosensor. However, there are some limiting factors. If we consider the Michaelis–Menten equation (see Section 3.5.2 above), the rate of reaction is directly proportional to the enzyme concentration, as follows:

$$v = \frac{k[E_0][S]}{K_m + [S]}$$

Provided there is sufficient enzyme present so that this process is not rate-limiting, there is no problem. However, if there is too much enzyme, or if the quality of the enzyme preparation is poor, so that considerable material is needed to provide sufficient units of enzyme activity (U), the excess of material can affect the rates of mass transport (principally diffusion) to the transducer. This factor is seldom mentioned in publications about biosensors. An example is given in Table 4.3 for urea, showing that trebling of the amount of enzyme from 25 to 75 U of urease caused a dramatic improvement in lifetime from three weeks to four months, with a slight lengthening of the response time and a slight deterioration in the detection limit. However, in general, there is little published systematic information on this aspect.

4.8.2 Immobilization Method

This has been discussed above in Chapter 3. To summarize, chemical (covalent bonding and cross-linking) methods result in longer lifetimes, but can limit the response by blocking the mass transport processes, and vice versa for physical methods. However, the chemical method may sometimes damage the enzyme, so causing a further diminution in response. This is counterbalanced by the more rapid loss of enzyme which results from the weaker bonding which is present in physical methods (entrapment or adsorption).

4.8.3 pH of Buffer

It is normally necessary to control the pH of the test solution fairly carefully. Commonly, a phosphate buffer at pH 7.4 is used. However, the optimum pH is very much dependent on the electron-transfer mediator being used. Wilson and Turner (1992) have provided a survey for glucose oxidase (GOD), and showed that there are three groups of mediators, depending on the pH, as follows:

- pH optimum, 5.6 (citrate buffer) – quinones and oxygen
- pH optimum, 7.5 (phosphate buffer) – diamines, ferrocenes and TTF–TCNQ
- pH < 4 – [$Fe(CN)_6^{3-}$] and indophenols

SAQ 4.6

List selective elements for the determination of the following:

(a) H^+ ions;
(b) F^- ions;
(c) Glucose;
(d) Oestradiol;
(e) Narcotic drugs.

Summary

This chapter describes the various criteria needed to assess the performance of a sensor. The essential characteristic of a sensor is its selectivity – already discussed earlier in Chapter 3. The selectivity features of ion-selective electrodes, including the use of the Nicholskii–Eisenman relationship, are presented. The range, linear range, and detection limit are described, and various time-related criteria, including the response time, recovery time and lifetime are discussed. The importance of accuracy, precision and reproducibility are outlined. These factors are exemplified by application to potentiometric and amperometric sensors and biosensors, including those used for urea, glucose, uric acid and amino acids. The effects of enzyme amount, immobilization method, transducer and pH on these criteria are also presented.

Further Reading

Arnold, M. A. and Reichnitz, G. A., 'Biosensors based on plant and animal tissues', in *Biosensors: Fundamentals and Applications*, Turner, A. P. F., Karube, I. and Wilson, G. S. (Eds), Oxford University Press, Oxford, UK, 1987, pp. 30–59.

Carey, W. P., 'Multivariate sensor arrays as industrial environmental monitoring systems', *Trends Anal. Chem.*, **13**, 210–218 (1994).

Gibson, T. D. and Woodward, J. R., 'Protein stabilization in biosensor systems', in *Biosensors and Chemical Sensors*, Eldham, P. G. and Wang, J. (Eds), American Chemical Society, Washington, DC, 1992, pp. 40–55.

Gibson, T. D., Hulbert, J. N., Parker, S. M., Woodward, J. R. and Higgins, I. J., 'Extended shelf life of enzyme-based biosensors using a novel stabilization system', *Biosensors Bioelectron.*, **7**, 701–708 (1992).

Kuan, S. S. and Guilbault, G. G., 'Ion selective electrodes and biosensors based on ISEs', in *Biosensors: Fundamentals and Applications*, Turner, A. P. F., Karube, I. and Wilson, G. S. (Eds), Oxford University Press, Oxford, UK, 1987, pp. 135–152.

Scheller, F. W., Pfeiffer, D., Schubert, F., Renneberg, R. and Kirsten, D., 'Application of enzyme-based amperometric biosenors to the analysis of "real" samples', in *Biosensors: Fundamentals and Applications*, Turner, A. P. F., Karube, I. and Wilson, G. S. (Eds), Oxford University Press, Oxford, UK, 1987, pp. 315–346.

Weber, S. G. and Webers, A., 'Biosensor calibration. *In situ* recalibration of competitive binding sensors', *Anal. Chem.*, **65**, 223–230 (1993).

Wilson, R. and Turner A. P. F., 'Glucose oxidase: an ideal enzyme', *Biosensors Bioelectron.* **7**, 165–185 (1992).

Chapter 5
Electrochemical Sensors and Biosensors

Learning Objectives

- To appreciate the operation and application of potentiometry to ion-selective electrodes.
- To apply the concept of activities to the determination of concentrations and interference effects.
- To understand how potentiometry can be used in gas sensors.
- To know how these concepts are applied to biosensors.
- To appreciate the development of the three generations of amperometric biosensors.
- To know how mediators are used in biosensors.
- To acquire a knowledge of a wide range of different amperometric biosensors.
- To understand the limitations and uses of conductivity measurements in gas sensors and biosensors.
- To know how field-effect transistors are used in a variety of sensors.

5.1 Potentiometric Sensors – Ion-Selective Electrodes

5.1.1 Concentrations and Activities

Ion-selective electrodes (introduced earlier in Chapter 2) are based on the principle of concentration cells, i.e. electrochemical cells linked by a membrane and containing the same half-cell electrode in each half of the cell, and differing only in the concentration of the analyte (as shown above in Figure 2.2). The membrane is *selective*, i.e. it responds to the analyte ion more than to other ions. The relationship between the emf of the cell and the analyte concentration is derived from the Nernst equation (See Section 2.2.3 above) and can be expressed in the following general form:

$$E = K + S \log [\text{ion}]$$

where E is the emf of the cell, S is the slope of the calibration graph (ideally 59.1 mV per decade of concentration), and [ion] is the concentration of the ion. Strictly speaking, the latter parameter should be the 'activity' of the ion (a_i), which gives the true thermodynamic Nernstian response. The activity is related to the concentration by the activity coefficient, γ, so that:

$$a_i = \gamma[\text{ion}]$$

where γ can be calculated from the Debye–Huckel theory, which estimates the effects of interaction between ions in a solution. The Debye–Huckel equation is given as follows:

$$-\log \gamma_i = (Az_i^2\sqrt{I})/(1 + Ba\sqrt{I}) \quad (5.1)$$

where A and B are constants arising from the theory, with values of 0.51 and 3.3×10^7, respectively, at 298 K, a is the ion size parameter (see Table 5.1), and z is the charge on the ion. The ionic strength, I, is a measure of the total ions in solution, weighted according to their charges, as in the following equation:

$$I = 1/2\Sigma[\text{ion}]_{i,j}z_{ij}^2$$

Table 5.1 Ion sizes for use in the Debye–Huckel equation

Ion	Size parameter (pm)
Sn^{4+}, Ce^{4+}	1100
H^+, Al^{3+}, Fe^{3+}, Cr^{3+}	900
Mg^{2+}	800
Li^+, Ca^{2+}, Cu^{2+}, Zn^{2+}, Sn^{2+}, Fe^{2+}	600
Sr^{2+}, Cd^{2+}, Hg^{2+}, S^{2-}, OAc^-	500
Na^+, Pb^{2+}, $CO_3{}^{2-}$, $SO_4{}^{2-}$, $HPO_4{}^-$	400
K^+, Ag^+, $NH_4{}^+$	300
Cl^-, F^-, Br^-, I^-, OH^-, $NO_3{}^-$, SH^-, $ClO_4{}^-$	300

DQ 5.1

Calculate the activity of the sodium and sulfate ions in a 0.01 M solution of sodium sulfate.

Answer

The ionic strength of a 0.01 M solution of sodium sulfate is given by:

$$\mathrm{I} = 1/2\{(0.02 \times 1^2) + [0.01 \times (-2)^2]\} = 0.03\ M$$

while the activity coefficients can be calculated as follows:

$$-\log \gamma(Na^+) = [0.51 \times (+1)^2 \times \surd 0.03]/[1 + (3.3 \times 10^7 \times 4 \times 10^{-8} \times \surd 0.03)] = 0.0719$$

and therefore $\gamma(Na^+) = 0.847$

$$-\log \gamma(SO_4{}^{2-}) = [0.51 \times (-2)^2 \times \surd 0.03]/[1 + (3.3 \times 10^7 \times 4 \times 10^{-8} \times \surd 0.03)] = 0.2876$$

and therefore $\gamma(SO_4{}^{2-}) = 0.516$

Hence, the activities are given as follows:

$$Na^+ = 0.847 \times 0.02\ M = 1.694 \times 10^{-2}\ M$$

$$SO_4{}^{2-} = 0.516 \times 0.01\ M = 5.16 \times 10^{-3}\ M$$

In practice, one can eliminate the effects of activity coefficients by making up all of the test solutions with a high concentration of the same ion which does not interfere (an ionic-strength adjuster*). Thus, the ionic strength is constant for each sample and so the activity coefficient is also constant. Therefore:*

$$\mathrm{E} = \mathrm{K} + \mathrm{S}\log \mathrm{a_i} \ \textit{becomes:}$$

$$\mathrm{E} = \mathrm{K} + \mathrm{S}\log(\gamma[ion])$$

giving:

$$\mathrm{E} = (\mathrm{K} - \mathrm{S}\log \gamma) + \mathrm{S}\log[ion]$$

This can be written as follows:

$$\mathrm{E} = \mathrm{K}' + \mathrm{S}\log[ion]$$

5.1.2 Calibration Graphs

When employing calibration graphs in the study of ISEs, the following important points should be noted:

1. The slope of a (calibration) graph is *Nernstian* if the slope, S, is $59.1/z$ mV ($\pm 1-2$ mV). Below this level of S, the slope is termed *sub-Nernstian* (the usual case), or is called *hyper-Nernstian* if greater (than $59.1/z$ mV). Improved performance may be achieved if the electrode is conditioned for 1–2 h in a solution of the ion of interest (ca. 0.01 M).

2. The linear range is usually between 10^{-5} and 10^{-1} M (depending on the ion), thus making ISEs suitable for many environmental and biological measurements.
3. Below 10^{-5} M, there may be curvature (see Figure 4.1 above), due to either approaching the detection limit or to the effect of an interferant.
4. Conditioning of the ISE should be carried out before preparation of calibration curves. To achieve this, the electrode is steeped in a 0.01 M solution of the ion to be analysed for 1–2 h, followed by 30 min in deionized water.
5. A criterion of stability for an ISE is generally that the cell potential does not vary by more than ±0.1 mV over a period of 60 s. At low concentrations, a more exact standard may be necessary, such as requiring that the potential should be stable within ±0.1 mV for 120 s. For example, a fluoride electrode may take 15–30 min to reach a steady-state condition at a concentration of 0.1 mg dm^{-3}.
6. The effect of interfering ions can be described by the Nicholskii–Eisenman equation, as discussed above in Sections 3.2.2 and 4.2.1.

SAQ 5.1

50 cm^3 of a solution of Cu(II) was analysed by using a multiple standard addition (potentiometric) method. When 1.00 cm^3 increments of 0.1 M Cu(II) were added to the test sample, the following readings were obtained:

Volume added (cm^3)	*E* (mV)
1.0	99.8
2.0	102.5
3.0	104.6
4.0	106.3
5.0	107.9

A blank solution gave a reading of 70.0 mV.
Estimate the concentration of copper in the original solution.

DQ 5.2

For a calcium ISE, the calibration slope, *S*, was +29.6 mV/decade in a 0.001 M solution. In a 0.001 M calcium chloride solution, the cell potential was −20.1 mV, while the potential in a solution containing a mixture of 0.001 M calcium chloride and 0.1 M sodium chloride was −19.8 mV.

Calculate the selectivity coefficient for calcium ions in the presence of sodium.

Answer

The ionic strengths of the analyte and mixed solutions are determined as in DQ 5.1 above.

$$I(CaCl_2) = 1/2\,(0.001 \times 2^2 + 0.002 \times 1^2) = 0.003$$

$$I(CaCl_2 + NaCl) = 1/2\,(0.001 \times 2^2 + 0.002 \times 1^2 + 0.1 \times 1^2 + 0.1 \times 1^2) = 0.103$$

The activity coefficient of Ca^{2+} in the analyte is given by:

$$-\log \gamma = 0.51 \times 2^2 \times \sqrt{(0.003)}/[1 + 3.3 \times 10^7 \times 6 \times 10^{-8}\sqrt{(0.003)}] = 0.100\,80$$

and so $\gamma(Ca^{2+}) = 0.793$.

Hence, the activity of Ca^{2+} in the analyte $= 0.001 \times 0.793 = 7.93 \times 10^{-4}$ M.

The activity coefficient of Ca^{2+} in the mixed solution is given by:

$$-\log \gamma = 0.51 \times 2^2 \times \sqrt{(0.103)}/[1 + 3.3 \times 10^7 \times 6 \times 10^{-8}\sqrt{(0.103)}] = 0.400$$

and so $\gamma(Ca^{2+}) = 0.3978$.

Hence, the activity of Ca^{2+} in the mixed solution $= 0.001 \times 0.3978 = 3.978 \times 10^{-4}$ M.

The activity coefficient of Na^+ in the mixed solution is given by:

$$-\log \gamma = 0.51 \times 1^2 \times \sqrt{(0.103)}/[1 + 3.3 \times 10^7 \times 4 \times 10^{-8}\sqrt{(0.103)}] = 0.114\,97$$

and so $\gamma(NaCl) = 0.7674$.

Hence, the activity of NaCl in the mixed solution $= 0.1 \times 0.7674 = 7.674 \times 10^{-2}$ M.

Therefore:

$$k_{Ca^{2+},Na^+} = [a_{Ca^{2+}}(a) \times 10^{(-19.8+20.1)/29.6} - a_{Ca^{2+}}(m)]/a(Na^+)^{n/z}$$

$$= [(7.929 \times 10^{-4} \times 1.0236) - (3.978 \times 10^{-4})]/(7.67 \times 10^{-2})^{2/1}$$

$$(as\ n = 2\,(Ca)\ and\ z = 1\,(Na))$$

$$= 0.000\,413\,8/0.005\,883 = 7.03 \times 10^{-2}$$

which shows that sodium interferes only weakly.

5.1.3 Examples of Ion-Selective Electrodes

5.1.3.1 Glass Membrane Type

The best, indeed almost universally known, example of an ISE is the glass membrane electrode for measuring hydrogen ion concentration or acidity, usually

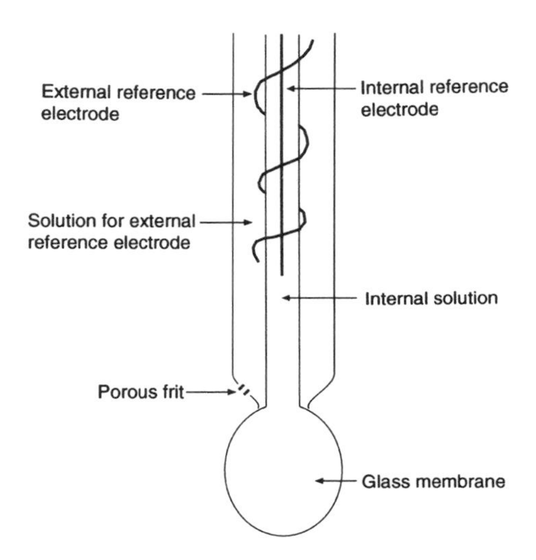

Figure 5.1 Schematic of a combination pH electrode. From Eggins, B. R., *Biosensors: An Introduction*, Copyright 1996. © John Wiley & Sons Limited. Reproduced with permission.

called the pH electrode. The thin glass membrane is highly selective to hydrogen ions over a very wide range of concentrations, with the composition of the glass being critical for this performance. If it is changed, this may make the glass membrane selective to other ions. The usual composition of the glass employed for detecting hydrogen ions is 22% Na_2O, 6% CaO, and 72% SiO_2. The basic reaction is as follows:

$$SiO^-Na^+ + H^+ = SiO^-H^+ + Na^+$$

$$E = K + 59.1 \text{ pH}$$

A typical pH combination electrode is shown in Figure 5.1. This type incorporates the second reference electrode in a concentric glass tube around the main electrode tube. Contact between this electrode and the test solution is through a small glass frit. The two reference electrodes are normally of the Ag/AgCl type. The hydrogen ion glass electrode is usually expressed as a pH electrode and calibrated in terms of pH rather than hydrogen ion activity, where:

$$\text{pH} = -\log a_{H+}$$

Therefore:

$$E = K + 0.059 \log a_{H+} = K - \text{pH}$$

and so:

$$\text{pH} = (K - E)/0.059$$

Other glass membrane ion-selective electrodes are available for measuring Na^+, Li^+, K^+ and Ag^+.

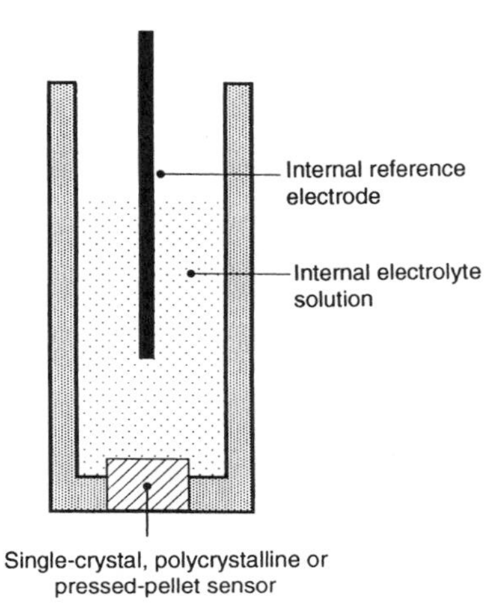

Figure 5.2 Schematic of a solid-state ion-selective electrode. From Eggins, B. R., *Biosensors: An Introduction*, Copyright 1996. © John Wiley & Sons Limited. Reproduced with permission.

5.1.3.2 Solid-State Type

The simple schematic in Figure 5.2 exemplifies the general structure for this type of electrode. Such a system normally has a separate reference electrode provided by the operator to dip into the test solution. It may be (but need not be) the same as the internal reference electrode built in by the manufacturer. The solid-state membrane can be a solid crystal, such as LaF_3 in the fluoride electrode, or a pressed pellet of powdered material, such as AgS in sulfide electrodes.

A single crystal of LaF_3 (doped with EuF_3) has been used in the fluoride ISE since 1966.

$$E = K - 59.1 \log a_{F^-}$$

The fluoride electrode is regularly used in water-treatment plants for measuring the fluoride levels in drinking water. However, most solid-state ISEs contain a pressed pellet of powdered material, such as silver sulfide in sulfide and silver electrodes. Examples of this type of ISE include Ag^+, Cl^-, Br^-, SCN^- and S^{2-}.

5.1.3.3 Liquid Ion-Exchange Membrane Type

The membrane is made of a hydrophobic material such as plasticized poly(vinyl chloride) (PVC). Absorbed into this membrane is the liquid ion-exchange material, such as vallinomycin (for potassium). In order to maintain the concentration level in the membrane, there is a reservoir of the ion-exchange liquid dissolved in an organic solvent. Figure 5.3 shows the details of this type of ISE, including the special reservoir for the ion-exchanger solution, as well as the reference solution

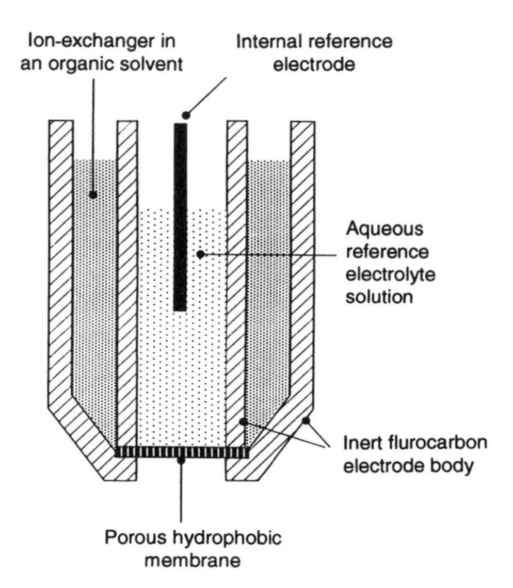

Figure 5.3 Schematic of a liquid ion-exchange membrane ion-selective electrode. From Eggins, B. R., *Biosensors: An Introduction*, Copyright 1996. © John Wiley & Sons Limited. Reproduced with permission.

and internal reference electrode. Some examples of this type of ISE are NO_3^-, Cu^{2+}, Cl^-, BF_4^-, ClO_4^- and K^+. The nitrate electrode is used extensively for the measurement of nitrate in soils and waters.

SAQ 5.2

A fluoride ion electrode is used to measure the fluoride concentration in a cup of tea. When immersed in a mixture of 25 cm^3 of tea and 25 cm^3 of an ionic-strength adjustment buffer, the electrode gave a reading of 98 mV. When 2.0 cm^3 of a 100 ppm fluoride solution was added to this mixture, the reading became 73 mV. Calculate the concentration of fluoride ions in the tea.

5.1.4 Gas Sensors – Gas-Sensing Electrodes

These are mainly based on pH electrodes and can detect gases which in aqueous systems form acidic or basic solutions. Here, a gas-permeable membrane is included in the arrangement, as shown in Figure 5.4. Between the membrane and the hydrogen-selective glass membrane is an internal electrolyte containing material that will form a buffer with the gas material. For example, for the ammonia electrode, ammonium chloride is used, so that an equilibrium is set up as follows:

$$NH_4Cl = NH_4^+ + Cl^-$$

$$NH_3 + H^+ = NH_4^+$$

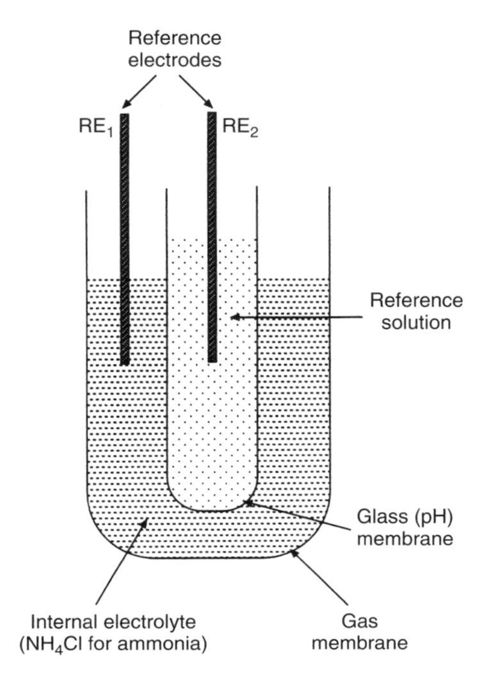

Figure 5.4 Schematic of a gas-permeable membrane electrode. From Eggins, B. R., *Biosensors: An Introduction*, Copyright 1996. © John Wiley & Sons Limited. Reproduced with permission.

so:

$$K = [NH_3][H^+]/[NH_4{}^+]$$

and therefore:

$$\log [NH_3] = pH + pK_a + \log [NH_4{}^+]$$

The presence of the high concentration of ammonium chloride keeps the concentration of ammonium ions constant. Hence, the logarithm of the ammonia concentration is directly proportional to the pH of the solution.

Electrodes for SO_2, NO_2, and H_2S are constructed in a similar way.

In biosensors, the most commonly used are H^+, $NH_4{}^+$ and NH_3 electrodes, which are all based on the pH principle. Occasionally, a CO_2, an I^-, or perhaps an S^{2-} electrode, may be used. Table 5.2 shows a selection of gas-sensing electrodes.

SAQ 5.3

How is selectivity achieved with electrochemical gas sensors?

Table 5.2 Some examples of dissolved-gas sensors

Gas	Inner solution	Sensor
CO_2	$NaHCO_3$	pH glass
SO_2	$NaHSO_3$	pH glass
HF	H^+	F–LaF_3
H_2S	pH 5 buffer	S^{2-}–Ag_2S
HCN	$KAg(CN)_2$	Ag^+–Ag
NH_3	NH_4Cl	pH glass

5.2 Potentiometric Biosensors

These are largely based on specific ion-selective or gas-selective electrodes.

5.2.1 pH-Linked

These are the simplest potentiometric biosensors, and are applicable to any system in which there is a change of pH during the (chemical) reaction. An appropriate enzyme must be immobilized on to the pH electrode to fabricate the sensor. There are many examples of such sensors, three of which are described in the following. The required enzyme is shown in the reaction scheme for each of these. As well as conventional pH electrodes, these types of biosensors are readily adapted for use with field-effect transistors (FETs).

5.2.1.1 Penicillin

$$\text{penicillin} \longrightarrow \text{penicilloate} + H^+$$

5.2.1.2 Glucose

Because gluconic acid is formed as a product, there is a change in the pH and so pH measurements can be used to monitor the reaction:

$$\text{glucose} + O_2 \xrightarrow{\text{GOD}} \text{gluconic acid} + H_2O_2$$

5.2.1.3 Urea

Urea was the first important analyte to be determined with a potentiometric biosensor (by Guillbault and Kuan in 1987). In this system, urea is hydrolysed by the use of the enzyme urease found in jack bean meal, as follows:

$$CO(NH_2)_2 + 2H_2O \xrightarrow{\text{urease}} 2NH_4{}^+ + CO_3{}^{2-}$$

The analysis may be carried out in a number of ways. With the aid of a suitable buffer, such as histidine, one can measure the reaction with a pH electrode and pH meter. The enzyme can be immobilized on to the pH electrode by using gelatin and glutaraldehyde. A simpler, although less reliable, method is to use a platinum electrode coated with polypyrrole instead of the standard glass pH electrode. Some other methods for the determination of urea are described below.

5.2.2 Ammonia-Linked

Any reaction in which ammonia is formed as a product can be monitored by using an ammonia-selective electrode.

5.2.2.1 Urea

In the reaction described above, we could use a cationic ammonium-selective electrode, or more commonly, we could make the solution alkaline and determine the liberated ammonia by the use of an ammonia-selective gas electrode. The latter has been the most successful method. Here, the urease is attached to the polypropylene membrane of an ammonia ISE. This has the highest sensitivity and the lowest detection limit (10^{-6} M), and can achieve 20 assays per hour with a relative standard deviation of $\pm 2.5\%$ over a range of 5×10^{-5} to 10^{-2} M.

5.2.2.2 Creatinine

$$\text{creatinine} \xrightarrow{\text{creatinase}} NH_3 + \text{creatine}$$

With the creatinase immobilized on the polypropylene membrane of an ammonia electrode, the latter was stable for 8 months and 200 assays, and had a detection limit of 8×10^{-6} M.

5.2.2.3 Phenylalanine

$$\text{L-phenylalanine} \xrightarrow{\text{phenylalanine ammonia-lyase}} NH_3 + \textit{trans}\text{-cinnamate}$$

This sensor is very highly selective, but has a poor range and slow response.

5.2.2.4 Adenosine

$$\text{adenosine} \xrightarrow{\text{adenine-deaminase}} NH_3 + \text{inosine}$$

In this system, the adenine–deaminase is cross-linked with glutaraldehyde on the ammonia electrode.

5.2.2.5 *Aspartame*

$$\text{aspartame} \xrightarrow{\text{L-aspartase}} NH_3 + C_6H_5CH_2CH(CO_2H)NHCOCHCHCO_2H$$

5.2.3 Carbon Dioxide-Linked

A few applications are known which involve the use of carbon dioxide gas ISEs in biosensors. These are described in the following.

5.2.3.1 *Urea*

In the reaction described above in Section 5.2.1.3, we could make the solution acidic and determine the liberated carbon dioxide with a carbon-dioxide-selective gas electrode.

5.2.3.2 *Oxalate*

The determination of oxalate in urine is important in the diagnosis of hyperoxaluria:

$$\text{oxalate} \xrightarrow{\text{oxalate decarboxylase}} CO_2 + \text{formate}$$

Phosphate and sulfate, which are usually present in urine, inhibit this enzyme, so oxalate oxidase can be used, although this enzyme is also inhibited by some anions:

$$\text{oxalate} \xrightarrow{\text{oxalate oxidase}} 2CO_2 + H_2O_2$$

5.2.3.3 *Digoxin*

Digoxin is immobilized on polystyrene beads, and a sample of digoxin is added together with a peroxidase-labelled antibody. The complexed peroxidase label is then reacted with pyrogallol and hydrogen peroxide and measured from the liberated carbon dioxide:

$$H_2O_2 + \text{pyrogallol} \xrightarrow{\text{peroxidase}} CO_2$$

SAQ 5.4

Compare the different types of urea biosensors.

5.2.4 Iodine-Selective

5.2.4.1 *Glucose*

The hydrogen peroxide formed from the reaction of glucose with glucose oxidase can be estimated by using it to oxidize iodide to iodine in the presence of peroxidase. The remaining iodide is measured by using the iodide-selective electrode:

$$\text{glucose} + O_2 \xrightarrow{\text{GOD}} \text{gluconic acid} + H_2O_2$$

$$H_2O_2 + 2I^- + 2H^+ \xrightarrow{\text{PO}} I_2 + 2H_2O$$

The iodide electrode follows the decrease in iodide concentration as it is consumed by the hydrogen peroxide.

5.2.4.2 Phenylalanine

L-Aminooxidase (LAO) and peroxidase (PO) are co-immobilized in a polyacrylamide gel on the surface of an iodide electrode. However, this sensor suffers more interference and selectivity problems than the ammonia-based sensor described above.

$$\text{L-phenylalanine} \xrightarrow{\text{LAO/PO}} H_2O_2$$

$$H_2O_2 + 2I^- + 2H^+ \longrightarrow I_2 + 2H_2O$$

5.2.4.3 Oestradiol

This is a potentiometric immunoassay. Anti-17β-oestradiol is immobilized on a gelatin membrane on the surface of an iodide ion-selective electrode. Peroxidase-labelled antigen and sample antigen are added (as shown in Figure 5.5(a)). The amount of sample antigen is inversely proportional to the labelled antigen, with the latter being determined by adding hydrogen peroxide and iodide, which is converted into iodine in the presence of the peroxidase. The remaining iodide is then determined by the ISE. A calibration graph of emf versus log [oestradiol] is obtained, as shown in Figure 5.5(b).

5.2.5 Silver Sulfide-Linked

5.2.5.1 Cysteine

$$Ag^+ + xR - S^{y-} \longrightarrow Ag(R{-}S)_x^{(1-xy)-}$$

The above reaction provides a direct potentiometric non-enzymatic method of analysis for cysteine, although it is not totally selective:

$$\text{cysteine} \longrightarrow \text{cystine (non-enzymatic)}$$

The following reaction:

$$\text{cysteine} + CN^- \xrightarrow{\beta\text{-cyanoalanine synthesase}} HS^- + \beta\text{-cyanoalanine}$$

is more specific, but in this case the cyanide interferes with the electrode.

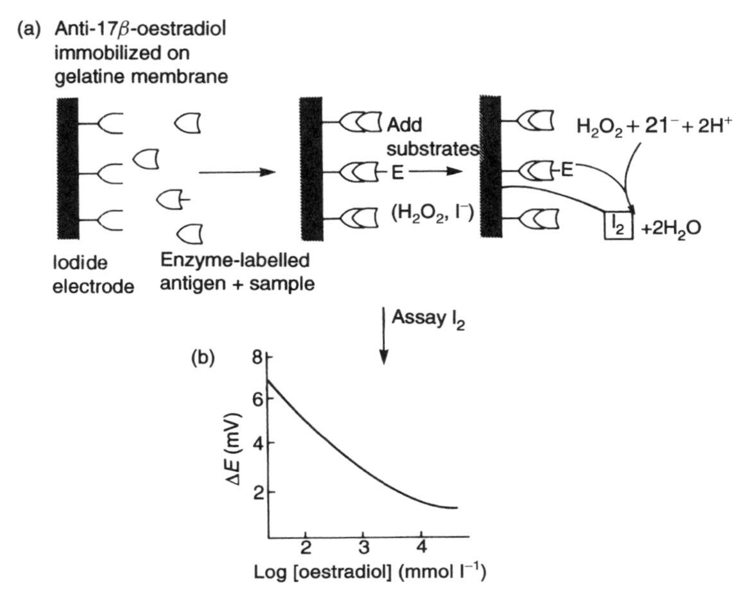

Figure 5.5 Determination of oestradiol-17β via an antibody-modified iodide electrode: (a) mechanism of the reaction; (b) the corresponding calibration graph of potential as a function of concentration. From Hall, E. A. H., *Biosensors*, Copyright 1990. © John Wiley & Sons Ltd. Reproduced with permission.

5.3 Amperometric Sensors

5.3.1 Direct Electrolytic Methods

On their own, voltammetric (amperometric) sensors do have some selectivity in that the reduction (oxidation) potential is characteristic of the species being analysed. As one sweeps the potential in the negative direction, the electroactive species are successively reduced (or oxidized if the sweep is in the positive direction). A typical differential pulse polarogram obtained by a linear sweep on a hanging-mercury-drop electrode for a mixture of six cations in 1 M HCl is shown in Figure 5.6.

Note that this method is not strictly a sensor technique – further details can be obtained from the AnTS text by Paul Monk, *Fundamentals of Electroanalytical Chemistry* (see the Bibliography). The selectivity is fairly limited, however, unless one uses modified electrodes (with extra selectivity 'incorporated into them'). Such applications were discussed briefly above in Chapter 3, and are considered in more detail below in Chapter 8, with application to a specific example.

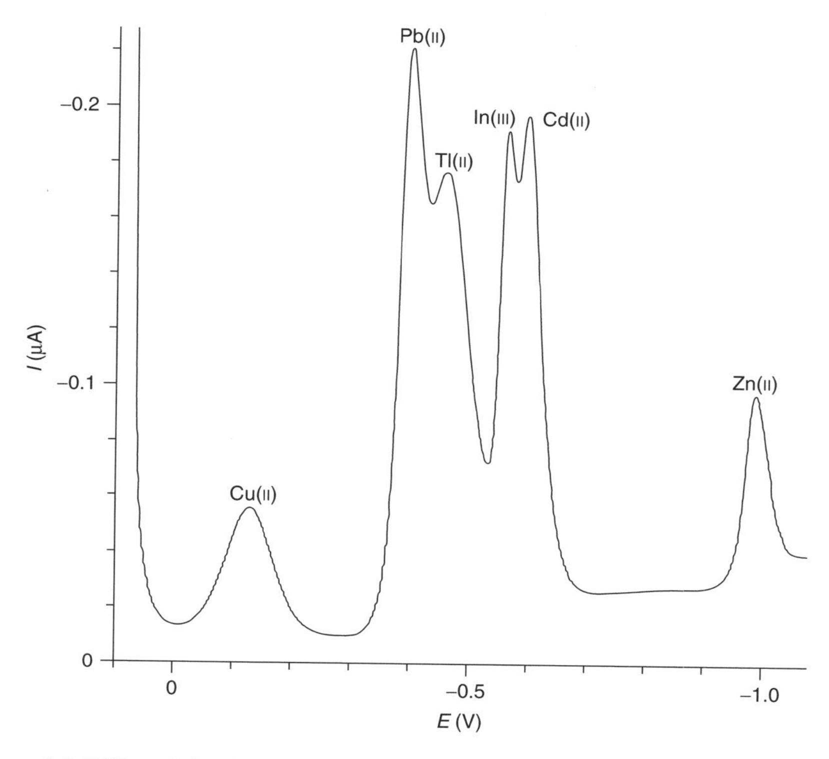

Figure 5.6 Differential pulse polarogram obtained for a mixture of six cations (Cu(II), Pb(II), Tl(II), In(III), Cd(II) and Zn(II)).

5.3.2 The Three Generations of Biosensors

Sometimes, the three modes of oxidation reactions that occur in biosensors are referred to as first-, second- and third-generation, as follows:

- First generation – oxygen electrode-based sensors
- Second generation – mediator-based sensors
- Third generation – directly coupled enzyme electrodes

However, there is some evidence that the mode of action of conducting-salt electrodes is really the same as that of a mediator, so that the third-generation description may not be strictly accurate.

SAQ 5.5

What is a mediator?

5.3.3 First Generation – The Oxygen Electrode

The original glucose enzyme electrode used molecular oxygen as the oxidizing agent, as follows:

$$\text{glucose} + O_2 \xrightarrow{\text{GOD}} \text{gluconic acid} + H_2O_2$$

The reaction is followed by measuring the decrease in oxygen concentration using a Clark oxygen electrode. This type of electrode was first developed in 1953 and uses the voltammetric principle of electrochemically reducing the oxygen, with the cell current being directly proportional to the oxygen concentration. The glucose oxidase is immobilized in polyacrylamide gel on a gas-permeable membrane covering the electrode, where the latter consists of a platinum cathode and a silver anode. Figure 5.7 shows a typical glucose sensor of this type. Such a system, in addition to being of great practical importance in the medical field, is also a useful model system on which many other biosensors designs can be based.

Several other biosensors have been developed which use oxidases and oxygen. A selection of some of these is given in Table 5.3.

Although these types of devices worked quite well, their operation raised a number of problems. First, the ambient level of oxygen needed to be controlled and constant, or otherwise the electrode response to the decrease in oxygen concentration would not be proportional to the decrease in glucose concentration.

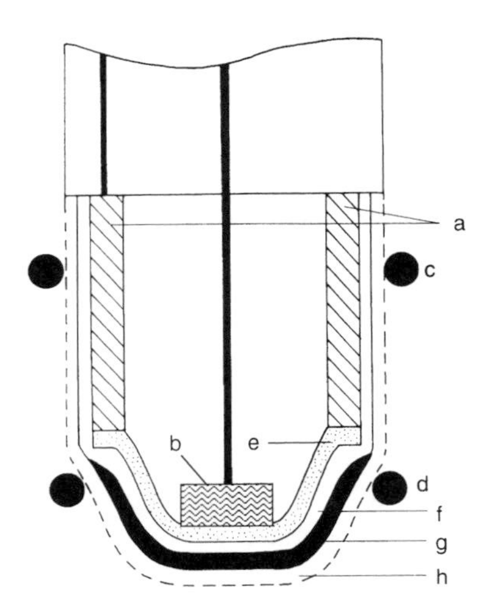

Figure 5.7 Schematic of the Clark-type glucose electrode, which uses two membranes: a, Ag anode: b, Pt cathode; c and d, rubber rings; e, electrolyte gel; f, 'Teflon' membrane; g, glucose oxidase on nylon net; h, cellophane membrane. From Hall, E. A. H., *Biosensors*, Copyright 1990. © John Wiley & Sons Ltd. Reproduced with permission.

Table 5.3 Some examples of oxidases which are used in biosensors. From Hall, E. A. H., *Biosensors*, Copyright 1990. © John Wiley & Sons Ltd. Reproduced with permission

Analyte	Enzyme	Response time (min)	Stability (days)
Glucose	Glucose oxidase	2	>30
Cholesterol	Cholesterol oxidase	3	7
Monoamines	Monoamine oxidase	4	14
Oxalate	Oxalate oxidase	4	60
Lactate	Lactate oxidase	—	—
Formaldehyde	Aldehyde oxidase	—	—
Ethanol	Alcohol oxidase	—	—
Glycollate	Glycollate oxidase	—	—
NADH	NADH oxidase	—	—

Another problem was that at the fairly high reduction potentials needed to reduce oxygen (−0.7 V):

$$O_2 + e^- \longrightarrow O_2{}^-$$

other materials might interfere.

The first way around this was to measure the oxidation of the hydrogen peroxide product:

$$H_2O_2 \longrightarrow 2H^+ + 2e^- + O_2$$

This was achieved by setting the electrode potential to +0.65 V. This is still fairly high in the opposite sense, and now the problem could be of interference from ascorbic acid, which is oxidized at this potential and is commonly present in biological samples.

A number of attempts have been made to regulate the oxygen level. Some of these were based on the fact that in the presence of the common enzyme, catalase, hydrogen peroxide is decomposed to water and oxygen, as follows:

$$H_2O_2 \xrightarrow{\text{catalase}} H_2O + O_2$$

However, only half the required oxygen is produced in this case, and then only if all of the hydrogen peroxide is recycled – in fact, only about 50% can be recycled. An alternative approach is to re-oxidize the water to oxygen at the anode:

$$H_2O - 2e^- \longrightarrow 2H^+ + O_2$$

Again, the standard electrode potential for this is very high at +1.23 V, and the application of such a potential would be likely to oxidize any interferants that might be present. Some success has been obtained with an oxygen-stabilized electrode in which a separate oxygen generation circuit is used, controlled through

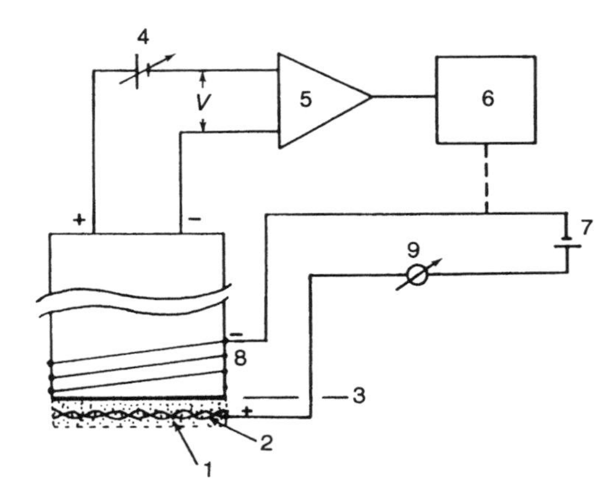

Figure 5.8 Schematic of the generation circuit used to obtain a constant oxygen concentration in an oxygen electrode: 1, immobilized enzymes; 2, platinum net; 3, 'Teflon' membrane of oxygen electrode; 4, reference voltage; 5, differential amplifier; 6, PID controller which regulates the current through the electrolysis circuit to maintain the differential voltage (V) at zero; 7, voltage source of electrolysis circuit; 8, platinum coil around electrode; 9, microammeter. Reprinted from **Biosensors: Fundamentals and Applications** edited by A. P. F. Turner, I. Karube and G. S. Wilson (1987), by permission of Oxford University Press.

a feedback amplifier from the analysing oxygen electrode. Such a system is shown in Figure 5.8.

The operational amplifier compares the measured current due to the presence of the oxygen with a standard potential. This is then fed back to control the electrolysis potential of the oxygen-generating circuit. The glucose oxidase is mixed with catalase and is embedded in a platinum-gauze electrode which functions as the anode of the generating circuit.

Another approach is to control the oxygen level by the rate at which an oxygen-containing buffer is pumped through the cell.

SAQ 5.6

Why is oxygen not a good mediator?

5.3.4 Second Generation – Mediators

An idea was developed to replace oxygen with other oxidizing agents – electron-transfer agents – which were reversible, had appropriate oxidation potentials and whose concentrations could be controlled. Transition-metal cations and their complexes were generally used for this purpose. Such materials are usually called

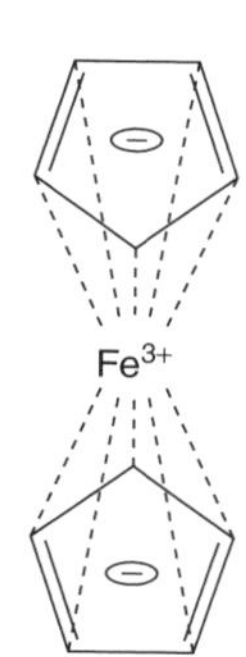

Figure 5.9 The structure of ferrocene. From Eggins, B. R., *Biosensors: An Introduction*, Copyright 1996. © John Wiley & Sons Limited. Reproduced with permission.

mediators. Many mediators are based on iron, either as ions or its complexes:

$$\text{Fe(III)} + e^- \longrightarrow \text{Fe(II)}$$

Free iron(III) ions do not make good mediators as they are subject to hydrolysis and precipitation as iron(III) hydroxide ($Fe(OH)_3$).

A common complex, which is sometimes used, is hexacyanoferrate(III), $[Fe(CN)_6]^{3-}$, formerly known as ferricyanide. However, the most successful mediators have been ferrocene complexes, (Fc) whose structures consist of a sandwich of the cation between two cyclopentadienyl (Cp) anions, as shown in Figure 5.9.

The following reactions apply for the various systems referred to above:

$$\begin{array}{ccc} Fe^{3+}{}_{aq} + e^- & \rightleftharpoons & Fe^{2+}{}_{aq} \quad (E^0 = +0.53\ \text{V}) \\ \downarrow H_2O & & \downarrow H_2O \\ Fe(OH)_3 + 3H^+ & & Fe(OH)_2 + 2H^+ \text{(hydrolysis)} \end{array} \tag{5.2a}$$

$$[Fe^{III}(CN)_6]^{3-} + e^- \rightleftharpoons [Fe^{II}(CN)_6]^{4-} \quad (E^0 = +0.45\ \text{V}) \tag{5.2b}$$

$$\underset{\text{ferrocene}}{[Fe^{III}(Cp)_2]^+} + e^- \rightleftharpoons Fe^{II}(Cp)_2 \quad (E^0 = +0.165\ \text{V};\ E_p(\text{Ox}) = +0.193\ \text{V};\ E_p(\text{R}) = +0.137\ \text{V}) \tag{5.2c}$$

Taking the example of glucose, the operation of a ferrocene-type mediator is as follows:

$$\text{glucose} + GOD_{Ox} \longrightarrow \text{gluconolactone} + GOD_R + 2H^+$$

$$GOD_R + 2Fc^+ \longrightarrow GOD_{Ox} + 2Fc$$

$$2Fc - 2e^- \longrightarrow 2Fc^+$$

The actual oxidation of the glucose is carried out by the flavin-adenine dinucleotide (FAD) component of the glucose oxidase, which is converted into $FADH_2$. The latter is re-oxidised to the FAD by the Fc^+ (mediator), followed by

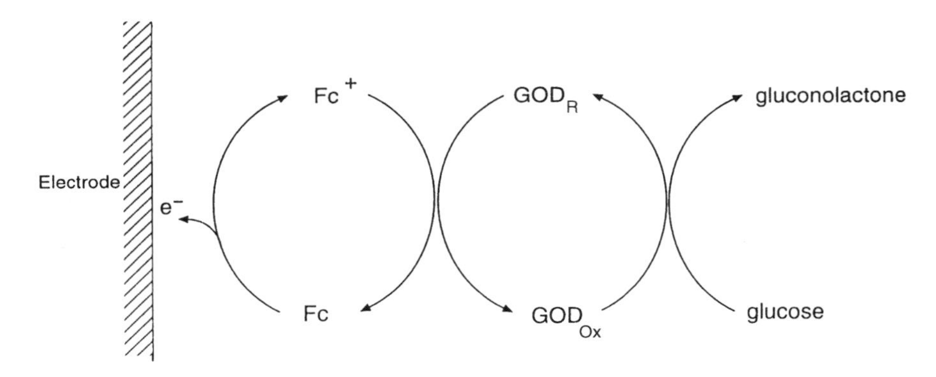

Figure 5.10 Mechanism of operation of a ferrocene-mediated biosensor for glucose: Fc, ferrocene; GOD, glucose oxidase. From Eggins, B. R., *Biosensors: An Introduction*, Copyright 1996. © John Wiley & Sons Limited. Reproduced with permission.

the re-oxidation of Fc to Fc^+ directly at an electrode, with the current flowing through the latter being an amperometric measure of the glucose concentration. This is illustrated in the cyclic reaction scheme shown in Figure 5.10.

DQ 5.3

What factors make good mediators?

Answer

The properties of a good mediator are as follows:

(i) It should react rapidly with the enzyme.
(ii) It should show reversible (i.e. fast) electron-transfer kinetics.
(iii) It should have a low over-potential for regeneration.
(iv) It should be independent of the pH.
(v) It should be stable in both its oxidized and reduced forms.
(vi) It should not react with oxygen.
(vii) It should be non-toxic.

Ferrocenes fit all of these criteria.

SAQ 5.7

Why is ferric sulfate a poor mediator?

As observed previously, an oxygen electrode is operated at −0.6 V, at which potential it is also likely to reduce ascorbic acid, which is normally present in large amounts in most enzyme or cell preparations.

Table 5.4 Redox potentials of some important reactions (at pH 7). From Eggins, B. R., *Biosensors: An Introduction*, Copyright 1996. © John Wiley & Sons Ltd. Reproduced with permission

Reaction	*E* (V)[a]	Reaction	*E* (V)[a]
Acetate–acetaldehyde	−0.60	Oxaloacetate–L-malate	−0.17
Acetone–propan-2-ol	−0.43	Ubiquinone–reduced ubiquinone	0.00
H^+–H_2	−0.42	Fumarate–succinate	+0.03
Xanthine–hypoxanthine	−0.37	Dehydroascorbate–ascorbate	+0.06
NAD^+–NADH	−0.32	Ferrocene–reduced ferrocene	+0.165
Oxidized–reduced glutathione	−0.23	O_2–H_2/O_2	+0.31
Cystine–cysteine	−0.22	$[Fe(CN)_6]^{3-}$/$[Fe(CN)_6]^{4-}$	+0.45
Acetaldehyde–ethanol	−0.20	Fe^{3+}–Fe^{2+}	+0.53
Pyruvate–L-malate	−0.19	O_2–H_2	+0.82

[a] Versus the standard hydrogen electrode (SHE).

Table 5.5 Redox potentials of some substituted ferrocenes and the corresponding electron-transfer rate constants when used as mediators in oxidation reactions involving glucose oxidase. From Eggins, B. R., *Biosensors: An Introduction*, Copyright 1996. John Wiley & Sons Ltd. Reproduced with permission

Derivative	*E* (V)[a]	k (10^5 dm^3 mol^{-1} s^{-1})
1,1′-Dimethyl	0.100	0.8
Acetic acid	0.142	—
Ferrocene[b]	0.165	0.3
Amidopentylamidopyrrole	0.200	2.07
Aminopropylpyrrole	0.215	0.75
Vinyl	0.253	0.3
Monocarboxylic acid	0.275	2.0
1,1′-Dicarboxylic acid	0.290	0.3
Methyltrimethylamino	0.387	5.3
Polyvinyl	0.435	—

[a] Versus the saturated-calomel electrode (SCE).
[b] Parent (unsubstituted) material.

Table 5.4 presents the redox potentials of some important reactions (at pH 7) – some of these will be discussed in the following sections.

The ring(s) of the cyclopentadienyl group may have various substituent groups attached. The presence of these groups affects the properties of the ferrocene, particularly the redox potential, and also the rate constant for electron transfer to the enzyme. Some examples are shown in Table 5.5.

The solubility is also affected, which is important in formulating the biosensor. Thus, 1,1′-dimethylferrocene is insoluble in water and has an E^0 of +0.1 V and a rate constant for reaction with glucose oxidase of 0.8×10^{-5} dm^3 mol^{-1} s^{-1},

Table 5.6 'Natural' and 'artificial' mediators and their redox potentials at pH 7. From Eggins, B. R., *Biosensors: An Introduction*, Copyright 1996. © John Wiley & Sons Ltd. Reproduced with permission

Natural	E (V)[a]	Artificial	E (V)[a]
Cytochrome a_3	+0.29	Hexacyanoferrate(III)	+0.45
Cytochrome c_3	+0.24	2,6-Dichlorophenol	+0.24
Ubiquinone	+0.10	Indophenol	+0.24
Cytochrome b	+0.08	Ferrocene	+0.17
Vitamin K_2	−0.03	Phenazine methosulfate	+0.07
Rubredoxin	−0.05	Methylene Blue	+0.04
Flavoproteins	−0.4 to +0.2	Phthalocyanine	−0.02
$FAD/FADH_2$	−0.23	Phenosafranine	−0.23
$FMN/FMNH_2$	−0.23	Benzylviologen	−0.36
$NAD^+/NADH$	−0.32	Methylviologen	−0.46
$NADP^+/NADPH$	−0.32		
Ferridoxin	−0.43		

[a] Versus the standard hydrogen electrode (SHE).

whereas ferrocene monocarboxylic acid is fairly soluble in water and has an E^0 of +0.275 V and a rate constant of 2.0×10^{-5} dm^3 mol^{-1} s^{-1}.

Many other suitable mediator materials are available, and can be classified into 'natural' and 'artificial' electron mediators. The former type includes molecules such as the cytochromes, ubiquinone, flavoproteins and ferridoxins, while artificial mediators include many dyestuffs, such as Methylene Blue, phthalocyanines and viologens. Table 5.6 presents a comparison of the redox potentials of a selection of these mediators, with the structures of some of them being shown in Figure 5.11.

5.3.4.1 Rate Constants

In general, we can write the rate mechanism as follows:

$$\mathrm{R} \rightleftharpoons \mathrm{Ox} + \mathrm{e}^-$$

$$\mathrm{E_R} + \mathrm{Ox} \xrightarrow{k_1} \mathrm{E_{Ox}} + \mathrm{R}$$

$$\mathrm{E_{Ox}} + \text{glucose} \xrightarrow{k_2} \mathrm{E_{red}} + \text{gluconolactone}$$

where E is an enzyme. If $k_1 < 10k_2/[\text{glucose}]$, then k_2 is fast and k_1 is the rate-determining step.

We can study the effect of mediators by cyclic voltammetry, and obtain an estimate of the rate constant. If we determine the cyclic voltammogram of a solution containing ferrocene monocarboxylic acid in a phosphate buffer (pH 7), which

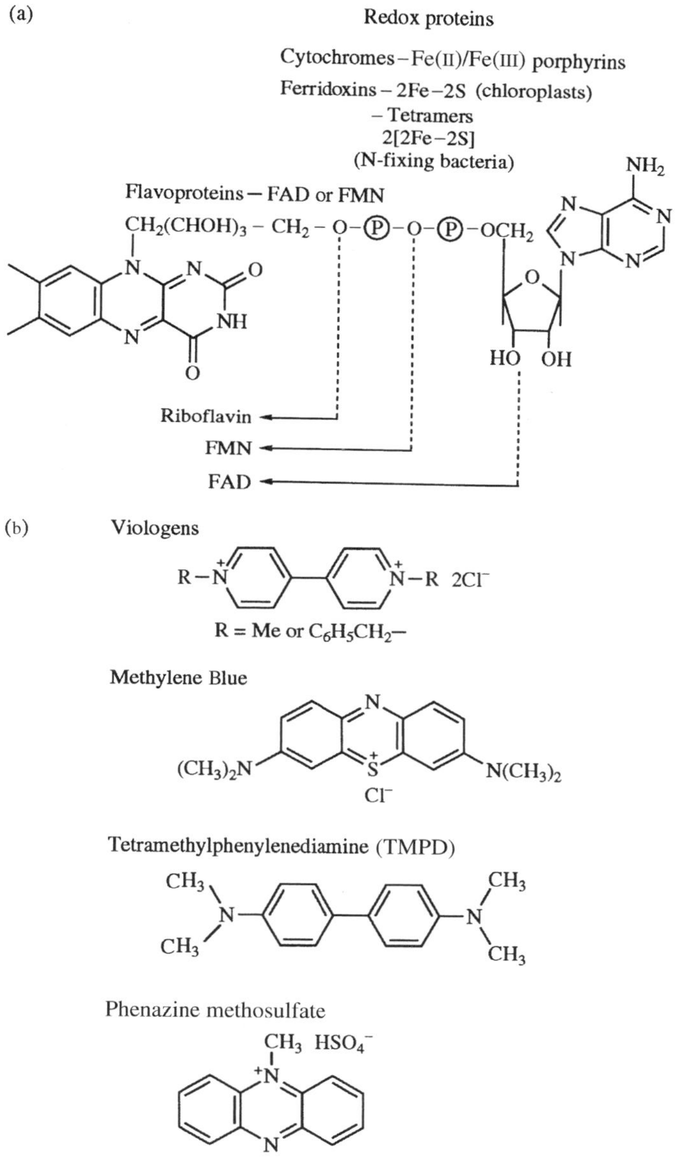

(a)

Redox proteins

Cytochromes – Fe(II)/Fe(III) porphyrins

Ferridoxins – 2Fe – 2S (chloroplasts)

– Tetramers

2[2Fe – 2S]

(N-fixing bacteria)

Flavoproteins – FAD or FMN

$CH_2(CHOH)_3 - CH_2 - O-Ⓟ-O-Ⓟ-OCH_2$

NH_2

HO OH

Riboflavin

FMN

FAD

(b) Viologens

$R-N^+ \ldots N^+-R\ 2Cl^-$

$R = Me$ or $C_6H_5CH_2-$

Methylene Blue

$(CH_3)_2N$ $N(CH_3)_2$ Cl^-

Tetramethylphenylenediamine (TMPD)

Phenazine methosulfate

$CH_3\ HSO_4^-$

Figure 5.11 Some examples of (a) natural and (b) artificial mediators used in oxidation reactions. From Eggins, B. R., *Biosensors: An Introduction*, Copyright 1996. © John Wiley & Sons Limited. Reproduced with permission.

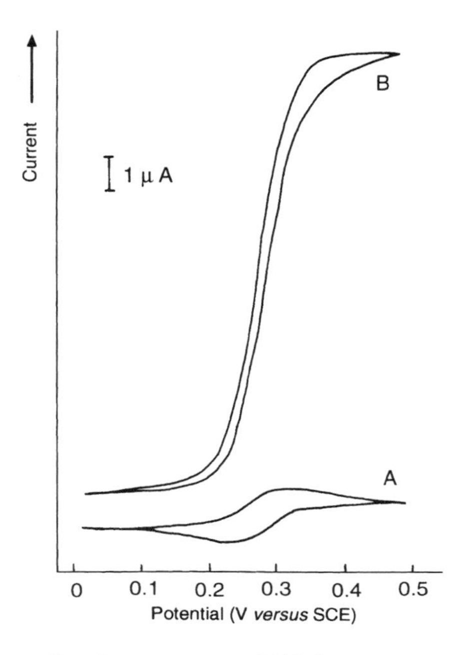

Figure 5.12 Catalytic cyclic voltammograms of (A) ferrocene monocarboxylic acid in the presence of glucose, and (B) the same system, but with the addition of glucose oxidase. Reprinted with permission from Cass, A. E. G., Davies, G., Francis, G. D., Hill, H. A. O., Aston, W. J., Higgins, I. J., Plotkin, E. V., Scott, L. D. and Turner, A. P. F., *Anal. Chem.*, **56**, 6567–6571 (1984). Copyright (1984) American Chemical Society.

also contains glucose, we obtain the typical reversible shape of the ferrocene cyclic voltammogram, as shown by curve A in Figure 5.12. If we now add glucose oxidase to this solution, we then obtain a catalytic plot with a greatly enhanced reduction peak and no oxidation peak (curve B) in Figure 5.12.

The rate constant is proportional to the relative height of the catalytic plot, so that we have:

$$i_k/i_d = f\,[\log(k_1)/v]$$

where v is the sweep rate.

The above relationship is shown in a graph of i_k/i_d versus $(k_f/a)^{1/2}$, where $a = nFv/RT$, i_k is the catalytic current (with GOD), i_d is the diffusion-controlled current, and k_f is the rate constant (Figure 5.13).

5.3.4.2 Formation of Biosensors Using Mediators

There are many ways in which mediators can be incorporated into biosensors. In the experiment described above, the components are all in solution. In a biosensor, both the enzyme and the ferrocene must be immobilized on the electrode.

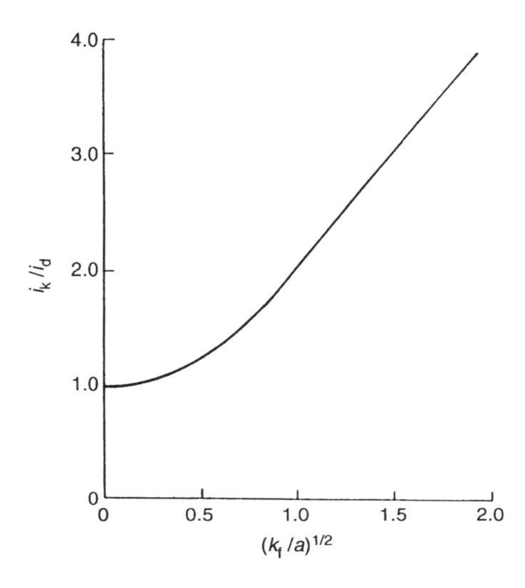

Figure 5.13 Theoretical plot of the ratio of the kinetic- to diffusion-controlled peak currents, i_k/i_d, as a function of the kinetic parameter, $(k_f/a)^{1/2}$. Reprinted from **Biosensors: Fundamentals and Applications** edited by A. P. F. Turner, I. Karube and G. S. Wilson (1987), by permission of Oxford University Press.

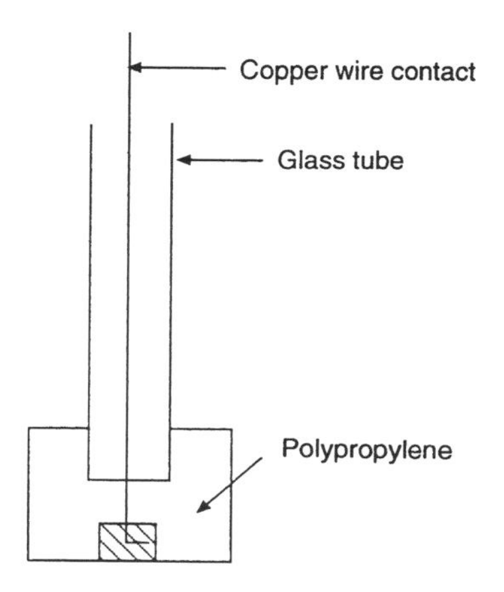

Figure 5.14 Schematic of a carbon paste electrode, for use with a mediator for biosensor applications. From Eggins, B. R., *Biosensors: An Introduction*, Copyright 1996. © John Wiley & Sons Limited. Reproduced with permission.

The simplest approach is to mix the mediator with carbon paste (liquid paraffin mixed with graphite powder) in a carbon-paste electrode, after which the enzyme is adsorbed on the surface and held in place with a membrane (as shown in Figure 5.14).

A more sophisticated approach was used by Cass and co-workers (1984). In this, graphite foil, with the edge plane exposed, was coated with dimethylferrocene by evaporation from a toluene solution. Glucose oxidase in a buffer was then immobilized on the surface by reaction with 1-cyclohexyl-3-(2-morpholinoethyl)carbodiimide-*p*-methyltoluenesulfonate. The sensor was then covered with a 'Nuclepore' membrane.

SAQ 5.8

Name three 'natural' and three 'artificial' mediators.

5.3.5 Third Generation – Directly Coupled Enzyme Electrodes

It may seem strange that a mediator is needed to couple an enzyme to an electrode. Why would it not be possible to reduce (oxidize) an enzyme directly at an electrode? The problem is that proteins tend to be denatured on electrode surfaces. In addition, the electron-transfer reaction may be slow and irreversible and hence requires an excessively high over-potential.

A possible approach is to modify the surface, e.g. with 4,4′-bipyridyl on a gold electrode. The bipyridyl is not itself electroactive, and nor is it a mediator. It forms weak hydrogen bonds with lysine residues on the enzyme, with such binding being temporary.

A better solution was developed by Albery and Cranston (1987) and Bartlett (1987), using organic-conducting-salt electrodes. In this system, tetrathiafulvalene (TTF) is reversibly oxidized, while tetracyanoquinodimethane (TCNQ) is similarly reversibly reduced (see Figure 5.15). A pair of these molecules form a *charge-transfer complex*, and it has been found that when such complexes are incorporated into an electrode, the surface becomes highly reversible and stable to many enzymes. Another important molecule for such an application is *N*-methylphenothiazine (NMP), which is sometime preferred to TTF.

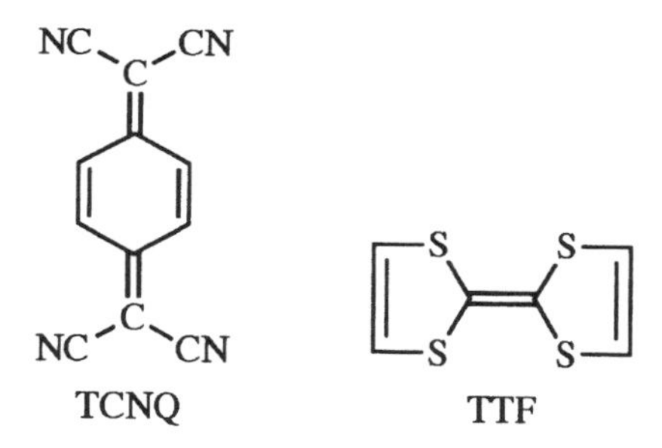

Figure 5.15 Structures of tetracyanoquinodimethane (TCNQ) and tetrathiafulvalene (TTF). From Eggins, B. R., *Biosensors: An Introduction*, Copyright 1996. © John Wiley & Sons Limited. Reproduced with permission.

These conducting salts can be built into electrodes in three ways, i.e. as single crystals, as pressed pellets or as a paste with graphite powder. These different forms vary slightly in their properties, in that the higher the crystalline form, then the better is the reversibility, but then, of course, the construction technique is more difficult.

5.3.5.1 Direct Enzyme–Electrode Coupling

Recently, immobilization techniques have been developed to 'wire' an enzyme directly to an electrode, thus facilitating rapid electron transfer and hence high current densities. In general, they involve an *in situ* polymerization process using a redox polymer. An example of this method used a glucose dehydrogenase (GDH) containing the redox centre, pyrroloquinolinequinone (PQQ), which was 'wired' to the glassy carbon electrode through a redox polymer, poly(vinyl pyridine), partially nitrogen-complexed with [osmium bis(bipyridine) chloride]$^{2+}$ and quaternized with bromoethylamine, (POs–EA), and cross-linked with poly(ethylene glycol 400 diglycidyl ether) (PEGDE). This oxygen-insensitive biosensor produced a very high current density of 1.8 mA cm^{-2} with 70 mM glucose, said to be three times higher than with a GOD sensor. The dissolved enzyme had a half-life of 5 days, but in continuous operation the current had decayed to the baseline in 8 h.

5.3.6 NADH/NAD$^+$

Nicotinamide – adenine dinucleotide (NAD) is a very common cofactor in many biochemical processes, coupling a hydrogen-transfer reaction with an enzyme reaction, as follows:

$$NAD^+ + RR'CHOH \longrightarrow NADH + RR'C{=}O + H^+$$

The structures of NADH and its redox form are shown in Figure 5.l6(a), with the corresponding reactions, i.e. reduction, oxidation and dimerization, that these species undergo given in Figure 5.16(b).

Unfortunately, the redox behaviour of NAD$^+$–NADH is rather irreversible at an electrode. The electrochemical reduction of NAD$^+$ does not give NADH, but leads to a dimer. NADH can be oxidized electrochemically to NAD$^+$, but at a substantial over-potential, i.e. above the standard redox potential. One can use a modified electrode, i.e. one coated with a suitable surface mediator, but such electrodes lack long-term stability. The use of conducting-salt electrodes can overcome this difficulty. NADH can be oxidized at −0.2 V (vs. Ag/AgC1) on an NMP$^+$–TCNQ$^-$ electrode. It can be used in a biosensor for lactate, with lactate dehydrogenase (LDH). The NADH was recycled on a glassy carbon (GC) electrode at +0.75 V, considerably above the standard potential of −0.32 V:

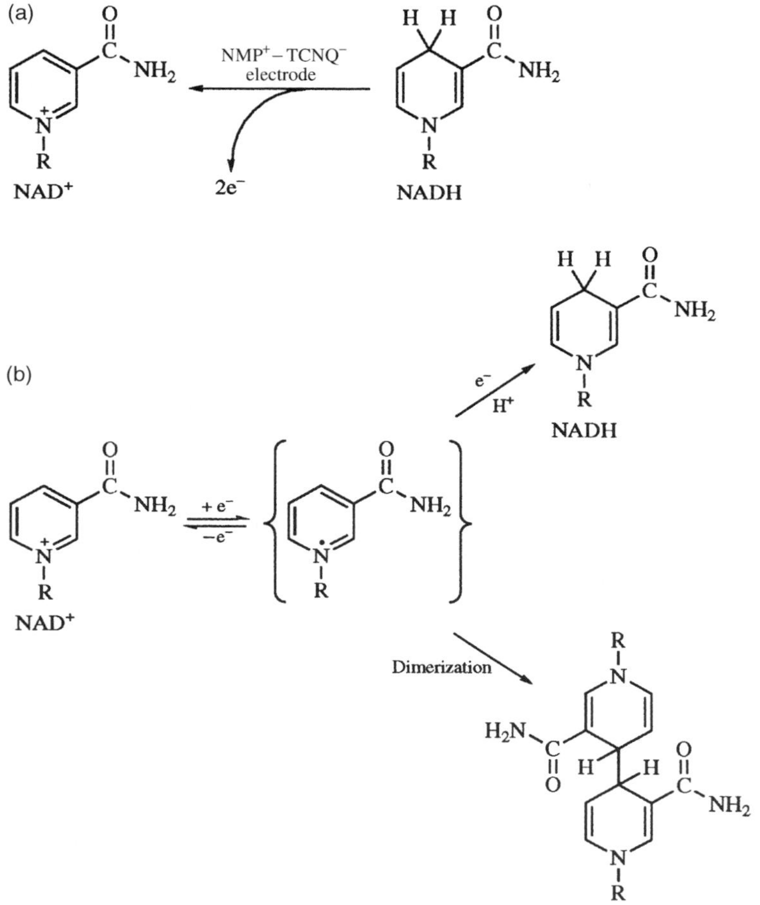

Figure 5.16 Structures (a) of nicotinamide–adenosine dinucleotide (NAD) and its reduced form (NADH), and (b) their corresponding reduction, oxidation and dimerization reactions. From Hall, E. A. H., *Biosensors*, Copyright 1990. © John Wiley & Sons Ltd. Reproduced with permission.

$$CH_3CHOHCO_2^- + NAD^+ \xrightarrow{\text{LDH}} CH_3COCO_2^- + NADH + H^+$$

$$NADH \longrightarrow NAD^+ + e^- + H^+ \ (\text{GC}, +0.75\ \text{V})$$
$$(NMP^+\text{–}TCNQ^-, -0.2\ \text{V})$$

There is a very large number of useful reactions that can be driven by such systems, particularly those involving dehydrogenases (estimated to be more than 250). Some of these are shown in Table 5.7. One problem, which has not yet been

Table 5.7 Some examples of NADH-coupled assays. From Eggins, B. R., *Biosensors: An Introduction*, Copyright 1996. © John Wiley & Sons Limited. Reproduced with permission

$\text{pyruvate} + \text{NADH} + \text{H}^+ \xrightarrow{\text{LDH}} \text{lactate} + \text{NAD}^+$
$\text{oxalacetate} + \text{NADH} + \text{H}^+ \xrightarrow{\text{MDH}} \text{malate} + \text{NAD}^+$
$\text{EtOH} + \text{NAD}^+ \xrightarrow{\text{ADH}} \text{CH}_3\text{CHO} + \text{NADH} + \text{H}^+$
$\text{G-6-P} + \text{NAD}^+ \xrightarrow{\text{G-6-PDH}} \text{glucono-6}'\text{-lactone-6-P} + \text{NADH} + \text{H}^+$
$\text{CH}_4 + \text{NAD}^+ + \text{H}_2\text{0} \xrightarrow{\text{MMO}} \text{CH}_3\text{OH} + \text{NADH} + \text{H}^+$
$\text{HCO}_2\text{H} + \text{NAD}^+ \xrightarrow{\text{FDH}} \text{CO}_2 + \text{NADH} + \text{H}^+$
$\text{NO}_3{}^- + \text{NADH} + \text{H}^+ \xrightarrow{\text{N(ate)R}} \text{NO}_2{}^- + \text{H}_2\text{O} + \text{NAD}^+$
$\text{NO}_2{}^- + 3\text{NADH} + 4\text{H}^+ \xrightarrow{\text{N(ite)R}} \text{NH}_3 + 2\text{H}_2\text{O} + 3\text{NAD}^+$
$2\text{Fe(CN)}_6{}^{3-} + \text{NADH} \xrightarrow{\text{diaphorase}} 2\text{Fe(CN)}_6{}^{4-} + \text{NAD}^+ + \text{H}^+$
$\text{MV}^+ + \text{NADH} \xrightarrow{\text{diaphorase}} \text{MV} + \text{NAD}^+$
$\text{androsterone} + \text{NAD}^+ \xrightarrow{\text{HSDH}} \text{5-androstane-3,17-dione} + \text{NADH}$
$\text{cholesterol} + \text{NAD}^+ \xrightarrow{\text{ChDH}} \text{cholestenone} + \text{NADH} + \text{H}^+$

overcome, is that NADH (and NAD^+) are expensive and are not very stable – nor can they easily be immobilized on a biosensor. Therefore, these materials have to be added directly to the analysis solution.

In addition to the lactate–pyruvate example mentioned above, ethanol can be readily measured by a biosensor based on this technique:

$$\text{C}_2\text{H}_3\text{OH} + \text{NAD}^+ \xrightarrow{\text{ADH}} \text{CH}_3\text{CHO} + \text{NADH} + \text{H}^+$$

Another important bioassay is for cholesterol:

$$\text{cholesterol} + \text{NAD}^+ \xrightarrow{\text{ChDH}} \text{cholestenone} + \text{NADH} + \text{H}^+$$

Other analytes that can be assayed in this way include L-amino acids, glycolic acid and, of course, NADH itself. While biosensors can be constructed for the analytes by using conducting-salt electrodes, a ferrocene-mediated electrode can

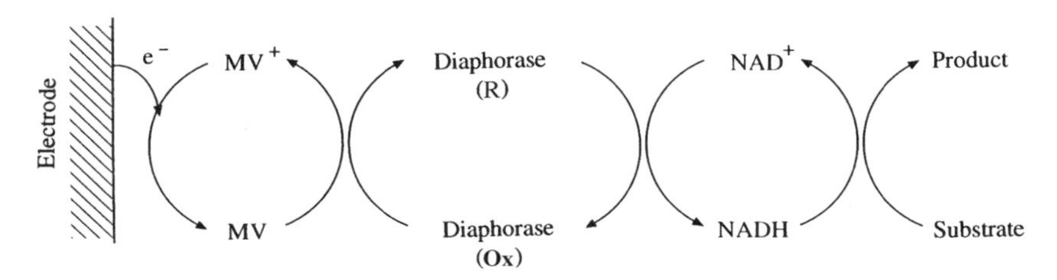

Figure 5.17 Reversible redox behaviour of NAD–NADH linked by diaphorase to a methyl viologen (MV) mediator. From Eggins, B. R., *Biosensors: An Introduction*, Copyright 1996. © John Wiley & Sons Limited. Reproduced with permission.

be used if the ferrocene is coupled to the NADH via the enzyme diaphorase (lipoamide dehydrogenase) (Figure 5.17).

SAQ 5.9

(a) What are the main problems with using NAD^+ as a mediator?
(b) How can these be overcome?
(c) Give three examples of the use of NAD^+ as a mediator.

5.3.7 Examples of Amperometric Biosensors

5.3.7.1 Glucose

It has been said that about half of the research papers published on biosensors are concerned with glucose. In addition to its metabolic and medical importance, this material provides a good standard compound on which to try out possible new biosensor techniques. There are, in fact, a number of different ways of determining glucose just by using electrochemical transducers. Reference to this has already been made earlier under other transducer modes (see Section 2.3.5 above).

Figure 5.18 shows the overall pattern. It can be seen that although there are several different ways in which glucose may be determined, glucose oxidase (GOD) lies at the centre of them all. The following points should be noted:

(i) Because gluconic acid is formed as a product, there is a change in pH and so measurement of the latter can be used to monitor the reaction:

$$\text{glucose} + O_2 \xrightarrow{\text{GOD}} \text{gluconic acid} + H_2O_2$$

(ii) The re-oxidation of the reduced form of GOD, directly at the electrode, is possible with special electrodes, as described in Section 5.3.5 above:

$$\text{glucose} + \text{GOD}_{\text{Ox}} \longrightarrow \text{gluconic acid} + \text{GOD}_{\text{R}}$$

$$\text{GOD}_{\text{R}} - 2e^- \longrightarrow \text{GOD}_{\text{Ox}}$$

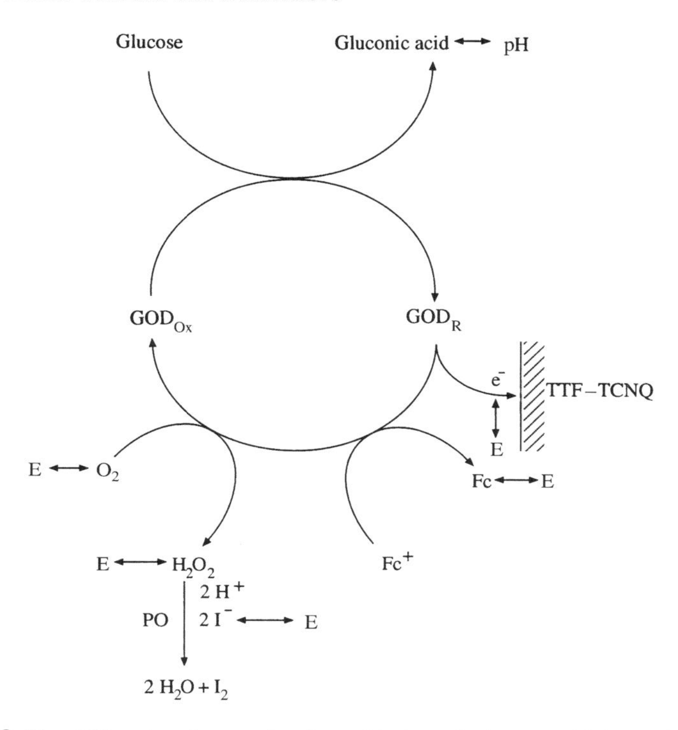

Figure 5.18 The different pathways for the oxidation of glucose which can be followed by the use of electrochemical sensors. From Hall, E. A. H., *Biosensors*, Copyright 1990. © John Wiley & Sons Ltd. Reproduced with permission.

(iii) As oxygen is consumed in the reaction, the decrease in oxygen concentration can be monitored with a Clark oxygen electrode, as described in Section 5.3.3 above:

$$\text{glucose} + O_2 \longrightarrow \text{gluconic acid} + H_2O_2$$

$$O_2 + 2e^- + 2H^+ \longrightarrow H_2O_2$$

(iv) An alternative is to monitor the hydrogen peroxide produced by the reduction of oxygen. This may be carried out directly by electro-oxidation at +0.6 V:

$$H_2O_2 - 2e^- \longrightarrow O_2 + 2H^+$$

(v) The hydrogen peroxide produced can be used to oxidize iodide to iodine in the presence of peroxidase (PO) and the decrease in iodide concentration measured with an iodide-selective electrode:

$$H_2O_2 + 2HI \xrightarrow{\text{PO}} H_2O + I_2$$

(vi) The oxygen may be replaced by a mediator, such as ferrocene (Fc), which can be detected by electrochemical oxidation:

$$\text{glucose} + 2\text{Fc}^+ \xrightarrow{\text{GOD}} \text{gluconic acid} + 2\text{Fc}$$

5.3.7.2 Lactate

Lactate ($CH_3CHOHCO_2H$) is an important analyte because of its involvement in muscle action, following which its concentration in blood rises. There are four different enzymes that can be used, with two of these being mediator-driven, while the other two are oxygen-driven. The processes in three of these cases lead to pyruvate (CH_3COCO_2H) and in the other to acetate, as shown in the following schemes:

(i) $\text{lactate} + \text{NAD}^+ \xrightarrow{\text{LDH}} \text{pyruvate} + \text{NADH} + \text{H}^+$

$$\text{NADH} \longrightarrow \text{NAD}^+ + 2\text{e}^- + \text{H}^+ \quad \text{(at the electrode)}$$

(ii) $\text{lactate} + 2[\text{Fe(CN)}_6]^{3-} \xrightarrow{\text{Cyt } b_2} \text{pyruvate} + 2\text{H}^+ + 2[\text{Fe(CN)}_6]^{4-}$

$$[\text{Fe(CN)}_6]^{4-} \longrightarrow [\text{Fe(CN)}_6]^{3-} + \text{e}^- \quad \text{(at the electrode)}$$

(iii) $\text{lactate} + \text{O}_2 \xrightarrow{\text{LOD}} \text{pyruvate} + \text{H}_2\text{O}_2$
(at the electrode; H_2O_2 may be oxidized or O_2 reduced)

(iv) $\text{lactate} + \text{O}_2 \xrightarrow{\text{LMO}} \text{acetate} + \text{CO}_2 + \text{H}_2\text{O}$
(at the electrode; O_2 is reduced)

(LDH, lactate dehydrogenase; LOD, lactate oxidase; LMO, lactate mono-oxidase; Cyt b_2, cytochrome b_2.)

5.3.7.3 Cholesterol

Too much cholesterol in the body is thought to be associated with heart disease, and therefore monitoring the level in the blood is becoming a routine health check analysis. The existing procedure is cumbersome and so a biosensor technique would be very useful. At the time of writing, no successful commercialization of such a biosensor has been achieved. However, a number of promising procedures have been developed in the laboratory.

Cholesterol is the most basic steroid alcohol, and is the major constituent of gallstones, from which it can be extracted. Cholesterol commonly occurs in the bloodstream as an ester. Therefore, any method of analysis involving the hydroxyl group needs a preliminary hydrolysis reaction step. This can be facilitated with the enzyme cholesterol esterase, which can be combined with the oxidizing enzyme cholesterol oxidase, which catalyses oxidation to cholestenone.

There are three possible approaches, all involving ferrocene as a mediator. The first of these involves coupling through NAD^+–NADH, diaphorase and then ferrocene, to an electrode, as shown in Figure 5.19.

The second approach uses oxygen, which is first converted into hydrogen peroxide–the latter is then coupled via peroxidase and ferrocene, as illustrated in Figure 5.20.

The third technique directly couples cholesterol oxidase to ferrocene. A novel method developed by Cassidy and co-workers (1993) uses a thin-layer cell

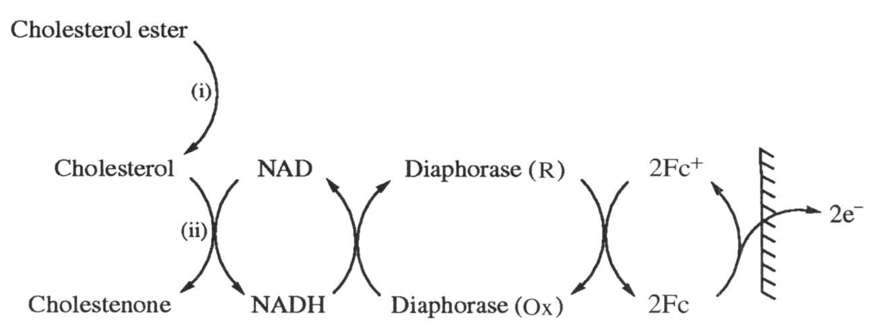

Figure 5.19 The reaction pathway of a biosensor which uses NAD–NADH, diaphorase and ferrocene for the analysis of cholesterol: (i) cholesterol esterase; (ii) cholesterol dehydrogenase. From Eggins, B. R., *Biosensors: An Introduction*, Copyright 1996. © John Wiley & Sons Limited. Reproduced with permission.

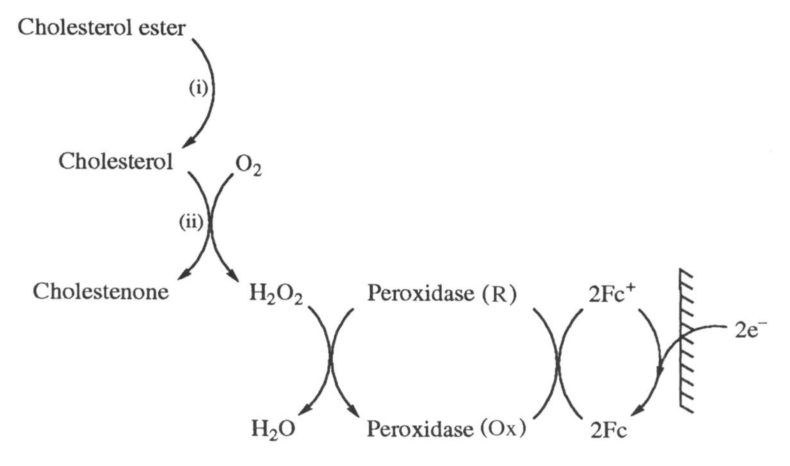

Figure 5.20 The reaction pathway of a biosensor which uses oxygen (converted into hydrogen peroxide), peroxidase and ferrocene for the analysis of cholesterol: (i) cholesterol esterase; (ii) cholesterol oxidase. From Eggins, B. R., *Biosensors: An Introduction*, Copyright 1996. © John Wiley & Sons Limited. Reproduced with permission.

containing $[Fe(CN)_6]^{3-}$, which shuttles oxygen backwards and forwards across the cell, thus setting up a steady-state situation. Such arrangements are sometimes referred to as 'fuel cells' in biosensor publications.

5.3.7.4 Phosphate

This is an interesting indirect assay which uses a glucose sensor. Glucose-6-phosphate is hydrolysed with acid phosphatase (AP) to free phosphoric acid and glucose, which can be determined with a glucose biosensor. However, phosphate inhibits the action of the phosphatase. The procedure then is to allow glucose-6-phosphate to react with phosphatase in the presence of the phosphate to be determined. This inhibits the reaction and so a reduced amount of glucose is formed:

$$\text{glucose-6-phosphate} \xrightarrow{\text{AP}} \text{glucose} + \text{phosphoric acid}$$

$$\text{glucose} + O_2 \xrightarrow{\text{GOD}} \text{gluconic acid} + H_2O_2$$

5.3.7.5 Starch

Starch is broken down by α-amylase to give dextrins and maltose. Glucoamylase will break down the maltose to glucose, which can then be determined with a glucose biosensor. The usual method in this case is to measure the hydrogen peroxide produced by the oxygen–glucose oxidase reaction, using the glucose at an electrode. However, there will also be glucose in the original solution and this must be filtered out – a double-membrane filter is used for this purpose. Everything can pass through the first membrane, and inside this are glucose oxidase, oxygen and catalase. The interfering glucose is broken down to hydrogen peroxide and gluconolactone, and the hydrogen peroxide is then oxidized by the catalase to oxygen. Hence, no free glucose or hydrogen peroxide will pass through the second membrane to the electrode, although maltose and oxygen will do so. Inside the second membrane are the glucoamylase and further glucose oxidase, which convert the maltose successively into glucose and then into gluconolactone and hydrogen peroxide, which is measured at the electrode. The reaction scheme for this process is illustrated in Figure 5.21.

5.3.7.6 Ethanol

This is an important target analyte because of the need to monitor blood-alcohol levels. Several methods have been developed for such an analysis. One approach employs a microbial sensor, which uses either *Acetobacter xyliniurn* or *Trichosporon brassicae*. Both of these bacteria catalyse the aerial oxidation of ethanol to acetic acid:

$$\text{ethanol} + O_2 \xrightarrow{\textit{A. xylinium}} \text{acetic acid} + H_2O$$

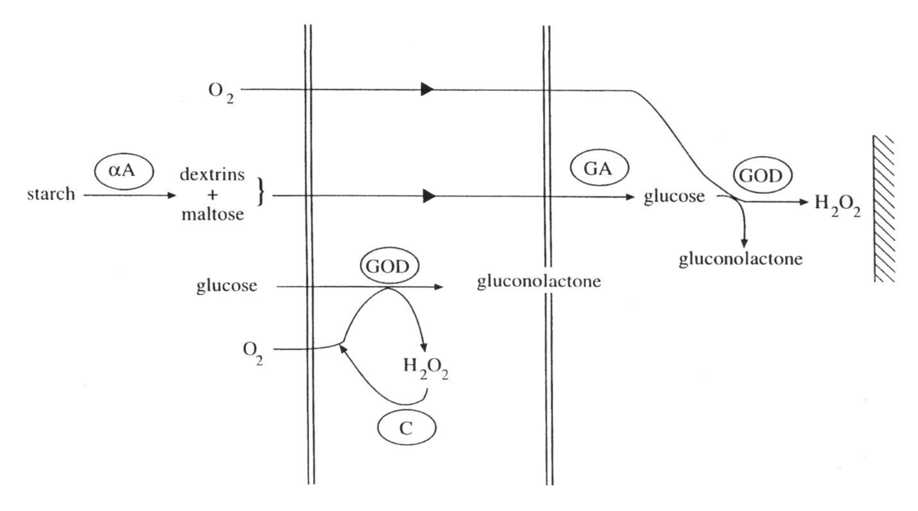

Figure 5.21 The reaction pathway for the determination of starch using a glucose-eliminating multilayer sensor: αA, α-amylase; GOD, glucose oxidase; C, catalase; GA, glucoamylase. Reprinted from **Biosensors: Fundamentals and Applications** edited by A. P. F. Turner, I. Karube and G. S. Wilson (1987), by permission of Oxford University Press.

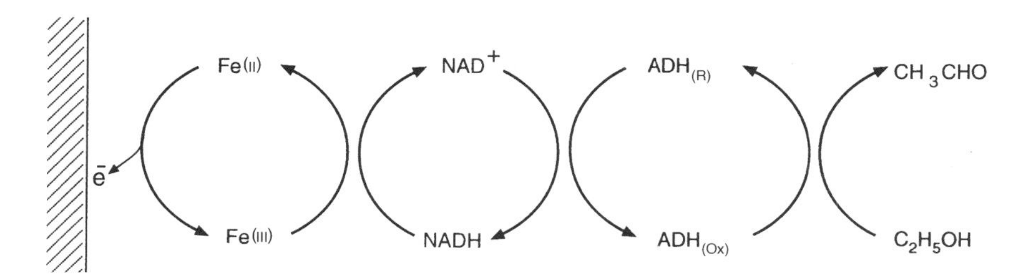

Figure 5.22 The reaction pathway of a biosensor which uses alcohol dehydrogenase (ADH) and nicotinamide–adenine dinucleotide (NAD), with an Fe(II)–Fe(III) mediator, for the analysis of ethanol. From Eggins, B. R., *Biosensors: An Introduction*, Copyright 1996. © John Wiley & Sons Limited. Reproduced with permission.

This reaction is followed by using an oxygen electrode. The method has been developed commercially in Japan, using membrane encapsulation, and works over the range 5–72 mM.

An alternative mediated method uses alcohol dehydrogenase (ADH), coupled via NAD^+–NADH and $[Fe(CN)_6]^{3-/4-}$ to an electrode (as shown in Figure 5.22). In addition, a bioluminescent detection method, also based on ADH and NAD^+, has been reported (see Section 2.7.5.2 earlier).

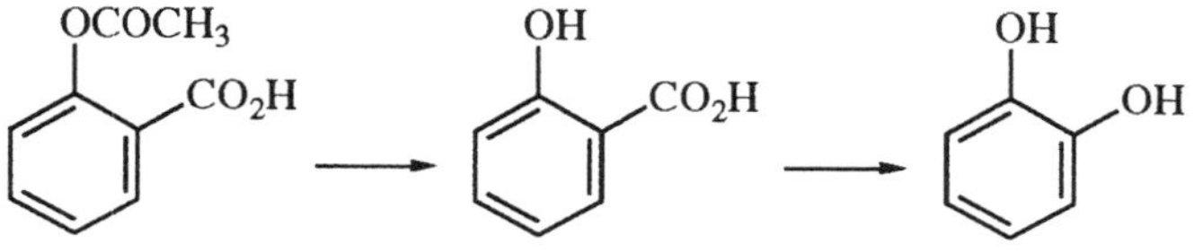

OCOCH$_3$ CO$_2$H → OH CO$_2$H → OH OH

Figure 5.23 The reaction scheme for the enzyme (salicylate hydroxylase)-catalysed hydrolysis and breakdown of aspirin. From Eggins, B. R., *Biosensors: An Introduction*, Copyright 1996. © John Wiley & Sons Limited. Reproduced with permission.

5.3.7.7 Aspirin

Levels above about 3 mM of aspirin in blood are toxic; the therapeutic dose is 1.1–2.2 mM. In blood, acetylsalicylic acid (aspirin) is converted into salicylic acid by hepatic esterases The conventional method for the determination of salicylic acid is spectrophotometric measurement of the complex formed with iron(III), which however, lacks specificity. An enzyme of bacterial origin, i.e. salicylate hydroxylase, catalyses the oxidation of salicylate to catechol via NADH (as shown in Figure 5.23). The catechol can be monitored by electro-oxidation using a screen-printed carbon electrode. However, this method still has some technical problems, which are currently being addressed in this author's group at the University of Ulster.

*5.3.7.8 Paracetamol (*N*-Acetyl-*p*-Aminophenol) ('Tylol')*

Poisoning by excess paracetamol, nowadays the commonest component of analgesic tablets, can cause irreversible liver damage. It is therefore vital that its identity and concentration in people who have taken overdoses be discovered as soon as possible. An efficient biosensor would be a very suitable device for achieving this. Paracetamol can be readily oxidized at a carbon paste electrode, but the latter, of course, is non-selective. However, the enzyme, aryl–acylamidase, will catalyse the hydrolysis of paracetamol to *p*-aminophenol, which is then electrochemically oxidizable, at a much lower potential, to quinoneimine, again using a disposable screen-printed carbon strip electrode. The reaction scheme for this process is shown in Figure 5.24.

SAQ 5.10

Compare the different biosensors used for the analysis of lactate.

5.3.8 Amperometric Gas Sensors

5.3.8.1 Oxygen

It is necessary to measure oxygen in a wide range of situations, including first-generation biosensors, oxidative metabolism in biochemistry, in medical situations, in the automotive industry, in water testing, and in the steel industry.

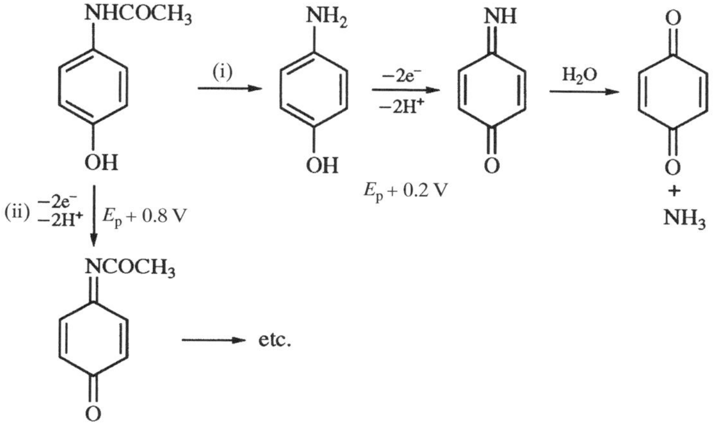

NHCOCH3
(i)
NH2
−2e−
−2H+
NH
H2O
O
OH
OH
O
O
Ep + 0.2 V
+
NH3
(ii) −2e− −2H+
Ep + 0.8 V
NCOCH3
etc.
O

Figure 5.24 The reaction scheme for the enzyme (aryl–acylamidase)-catalysed hydrolysis and oxidation of paracetamol (i), compared with that of direct oxidation using a carbon paste electrode (ii). From Eggins, B. R., *Biosensors: An Introduction*, Copyright 1996. © John Wiley & Sons Limited. Reproduced with permission.

The simplest type of amperometric sensor for oxygen is based on the polarographic principle and has been developed into the well-known Clark electrode (see Section 5.3.3 above). A commercial version of this for use in the automotive industry is shown in Figure 5.25. In this, a solid electrolyte ceramic made from ZrO_2/Y_2O_3 or CaO is coated with platinum electrodes. A potential is applied across the electrodes and the diffusion-limited current, which is directly proportional to the oxygen concentration, is measured.

Sulfur dioxide has a redox potential of about 750 mV, which is more cathodic than oxygen and so will not interfere. However, gases such as nitric oxide (NO) or chlorine, which have redox potentials more anodic than oxygen, could interfere,

Cathode
O2
A
E
Solid electrolyte (ZrO2/Y2O3)
O2
Anode

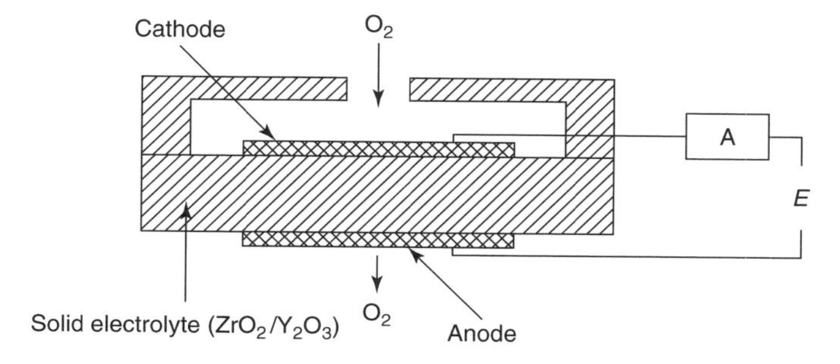

Figure 5.25 Schematic of a diffusion-controlled, limiting-current sensor. © R. W. Catterall 1997. Reprinted from **Chemical Sensors** by R. W. Catterall (1977) by permission of Oxford University Press.

although usually these are only present in trace amounts (except perhaps when chlorine is used to disinfect drinking water). Selective membranes may help to filter out interfering gases.

In medical applications, nitrous oxide or halothane can also be potential interferants.

5.3.8.2 Nitrous Oxide

Nitrous oxide is electrochemically reducible at −1.2 V on a silver electrode, whereas oxygen is reduced at less than −0.65 V:

$$N_2O + H_2O + 2e^- = N_2 + 2OH^-$$

A scheme has been devised for measuring both nitrous oxide and oxygen in the presence of each other, by applying successive polarizing pulses of −0.65 and −1.45 V. The current at −0.65 V is proportional to the oxygen concentration, while that at −1.45 V is proportional to the total gas concentration. Hence, by subtraction one can obtain the nitrous oxide concentration.

5.3.8.3 Halothane

The reduction potential of halothane on a silver electrode (−0.43 V) is too close to that of oxygen (−0.56 V) to use the same methods as for mixtures of O_2 and N_2O. However, at short times the reduction of halothane is very slow (under kinetic control), whereas oxygen is reduced under diffusion-controlled conditions at all times from < 20 to >50 ms. At longer times, the reduction of halothane becomes diffusion-controlled:

$$\mathrm{Ag + e^- + CHClBrCF_3 \xrightarrow{slow} (Ag\ldots Br\ldots CHClCF_3)^- \xrightarrow{fast\,+\,H_2O\,+\,e^-} Ag}$$
$$\mathrm{+\, Br^- + CH_2ClCF_3 + OH^-}$$

A single pulse of −1.45 V is applied and the current is sampled in two time zones at < 20 and >50 ms. Analysis of the data yields the concentrations of both oxygen and halothane in the mixture.

5.3.8.4 Biological Oxygen Demand

Biological oxygen demand (BOD) is one of the most important regular assays carried out on water. This represents a measure of how much oxygen is used up in oxidizing all of the biological organic matter in a water sample. The conventional method of determination requires an incubation period of up to 5 days. However, a biosensor based on the oxygen electrode would be ideal for this assay. The microorganisms, *Clostridium butyricum* and *Trichosporon cuteneum*, are suitable for such an application. These species are mounted in front of an oxygen electrode, and the latter is flushed with an oxygen-saturated buffer solution and the base oxygen current is read. The sample is then injected and after stabilization the oxygen current is read again. The difference in the two readings is linearly proportional to the result obtained by using conventional BOD analysis.

This type of sensor is designed to work in the temperature range 25–30°C. However, a BOD sensor has been developed for use at higher temperatures, employing thermophilic bacteria, isolated from hot springs – such a device is stable at temperatures of 60°C and above. This is useful for monitoring hot water outputs from factories.

5.3.8.5 Carbon Monoxide

Despite the disappearance of coal gas, poisoning and death caused by the emission of carbon monoxide from motor car exhaust fumes, domestic solid-fuel heater fumes or oil-heating fumes are unfortunately all too common. Existing devices for the detection of CO are expensive and not very selective. These are usually based on infrared spectrophotometric measurements. The problem is that CO is such a very simple molecule, and behaves very similarly to oxygen in many situations. However, some bacteria have been found, especially in anaerobic cultures, which contain enzymes that will catalyse the oxidation of CO to CO_2.

5.4 Conductometric Sensors and Biosensors

5.4.1 Chemiresistors

Sensors based purely on variation of the resistance of the device in the presence of the analyte generally lack selectivity. However, the use of an array of several sensing elements, each with a slightly different resistance response, can be used to detect quite complex mixtures. Each sensing element can be coated with a different conducting polymer, or made from sintered metal oxides. These arrays can develop a unique signature for each analyte. Such signatures can be obtained from mixtures of analytes and can be used to test flavourings in beers and lagers (and perhaps wines?) and to test the aromas of coffee blends. This type of system is often referred to as an 'electronic nose'. Evaluation of the signal output from an array of maybe 12 to 20 sensor elements makes use of *neural network analysis*, which simulates brain function activity. In this way, non-parametric, non-linear models of the array response can be constructed. A number of these devices have been commercialized.

5.4.2 Biosensors Based on Chemiresistors

If the sensing elements in such sensors is biological in nature (such as enzymes or antibodies), we can envisage another type of biosensor. Some examples of these are presented in the following:

(i) $$\text{Urea} + 2H_2O = 2NH_4^+ + HCO_3^-$$

This reaction clearly involves a change in the ions, and can be followed conductometrically with improved speed and sensitivity when compared to calorimetric methods. Conductance measurements are unaffected by colour or turbidity.

(ii) Many other enzymes result in appropriate changes in conductivity, for example:
- amidases – generate ionic groups
- dehydrogenases and decarboxlases – result in separation of unlike charges
- esterases – involve proton migration
- kinases – cause changes in the degree of ion association
- phosphatases and sulfatases – result in a change in size of the charge-carrying groups

5.4.3 Semiconducting Oxide Sensors

The conductivities of certain transition-metal oxides are affected by the adsorption of gas species on to their surfaces. The best examples of these are zinc oxide and tin dioxide (see Figure 2.32 earlier). Sensors based on this principle can be designed to detect carbon monoxide, carbon dioxide, ethanol and other organic vapours down to 1–50 ppm. This type of sensor is fabricated by using vapour deposition techniques. A 0.3 μm layer of palladium-doped tin oxide is grown on a silicon oxide layer on top of a ferrite substrate. On the other face of the ferrite is a thick layer of ruthenium oxide which acts as a heater to bring the sensor to its semiconducting operating temperature. The electrical resistance is measured between two gold contacts situated on the tin-oxide layer.

In such a system, n-type semiconductors are preferred because they give an electrical resistance change from high to low in the presence of the analyte gas. Some atmospheric oxygen is chemisorbed on the surface by reacting with the excess of electrons from the SnO_2 to give adsorbed oxygen anions (O^-_{ads}), which results in a low conductivity (high resistance). The adsorbed gas reacts with the chemisorbed oxygen anions, according to the following:

$$G + O^-_{ads} \longrightarrow GO^-_{des} + e^-$$

The gas removes the chemisorbed oxygen and is oxidized. Thus, gases which are reducing agents can be detected. The electrons produced cause an increase in the electrical conductivity of the SnO_2, which is measured as a decrease in resistance, and is proportional to the amount of gas present.

ZnO is an n-type semiconductor and can act in a similar way, although such sensors are somewhat lacking in selectivity. An improved selectivity can be obtained by varying the operating temperature and by the addition of other metal elements, such as platinum or palladium, which catalyse the oxidation of certain gases.

Sensors of this type are available commercially for the detection of flammable gases in buildings and homes. Another application is in the automotive industry.

For example, a titanium-dioxide-based sensor has been used to measure the oxygen content of exhaust gases. The resistance of the active layer increases with the amount of oxygen.

Semiconductor sensors also respond to changes in atmospheric humidity when water molecules are adsorbed on to the sensor's surface. However, the mode of action for this is not yet clear.

5.5 Applications of Field-Effect Transistor Sensors

DQ 5.4

How do field-effect transistors operate as sensors?

Answer

In order to convert the device into a sensor, the metal of the gate of an insulated-gate field-effect transistor (IGFET) is replaced by a chemically sensitive surface or membrane. These membranes are in contact with the analyte solution. A reference electrode completes the circuit via the gate-voltage bias. The membrane potential minus the solution potential corrects for the gate voltage V_g*, according to the following:*

$$\phi_{mem} - \phi_{sol} = 1/n_i F(\mu_{isol} - \mu_{i,mem})$$

$$1/n_i F(\mu_{isol} - \mu_{i,mem}) = E^0 + \frac{RT\, ln\, a_i}{n_i F}$$

and so:

$$\phi_{mem} - \phi_{sol} = E^0 + \frac{RT\, ln\, a_i}{n_i F}$$

$$I_D = K(V_G - V_T - E_{REF} - \frac{RT\, ln\, a_i}{n_i F} - V_D/2)V_D n_i F$$

where $K = C_0 W \mu_n / L$*, in which* C_0 *is the capacitance of the insulator,* W *is the width of the channel,* L *is the length of the channel, and* μ_n *is the electron mobility.*

The FET may be operated by measuring the drain current I_D *at constant* V_G*, or by measuring the gate voltage* V_G *at constant* I_D*. The circuits used for these two modes of measurement were described earlier in Chapter 2 (see Figures 2.27 and 2.28).*

5.5.1 Chemically Sensitive Field-Effect Transistors (CHEMFETs)

The simplest means of application is to use what is virtually a bare IGFET, with the gate consisting of a layer of palladium evaporated on to a silicon chip, and covered by a 100 nm oxide layer. This is highly specific for hydrogen gas down to 0.01 ppm: the response is given by $V = k_p$, with $k = 27$ mV per ppm. There is some sensitivity of this electrode to CO, NH_3, H_2S, CH_4, and C_4H_{10}.

With the addition of a layer of iridium, the sensitivity to ammonia is increased and that to hydrogen decreased.

5.5.2 Ion-Selective Field-Effect Transistors (ISFETs)

Ionophores are the most useful type of ion-selective (polymer) materials to use in FET devices, as is the case with ISEs. However, with FETs there are also some special responses for H^+. The first pH ISFET used the bare insulator gate as the ion-sensitive layer, although SiO_2 was not very effective, owing to its ease of hydroxylation. However, silicon nitride (Si_3N_4) gate devices are not hydrolysed and are highly selective to H^+ ions, with a response of 50–60 mV per decade (pH). TiO_2 and Ge show similar responses. These semiconductor materials can be handled by the same techniques as those used for FET chip fabrication. For other ions, such techniques are less successful. However, for Na^+ ions, borosilicate glass can be deposited in the gate region by using integrated circuit fabrication processes. Polymer membranes, incorporating valinomycin crown ethers, have been used successfully for K^+, and with *p*-(1,1,3,3-tetramethylbutyl)phenyl phosphoric acid for Ca^{2+}. The responses are < 40 mV per decade, unless the membrane is thicker than 100 μm.

A quadruple-function ISFET for H^+, Na^+, K^+ and Ca^{2+} has been described for clinical applications – this uses an Si_3N_4 bare gate for H^+, glass for Na^+, a phosphoric acid derivative (ionophore) for Ca^{2+} and valinomycin for K^+. This device worked satisfactorily under laboratory conditions, but under clinical conditions (whole blood samples) there were problems with the sodium analysis.

The use of heterogeneous membranes has met with some success, with the best of these being polyfluorinated phosphazine (PNF) mixed with silver salts, particularly silver chloride (75%) with PNF (25%). The response for Cl^- was 52 mV per decade. Changing the mixture and including Ag_2S or AgI can adjust the selectivity to favour a particular ion.

5.5.3 FET-Based Biosensors (ENFETs)

The use of enzyme-based FETs provides an excellent approach to producing miniaturized biosensors. Usually, a dual-gate arrangement is employed, as shown in Figure 5.26.

The pH-based FET is the most often used in this case, for example, with penicillin, glucose and urea. Indeed, a dual glucose–urea biosensor has been described, using three FET gates, with one as a reference, and the other two employing glucose oxidase and urease.

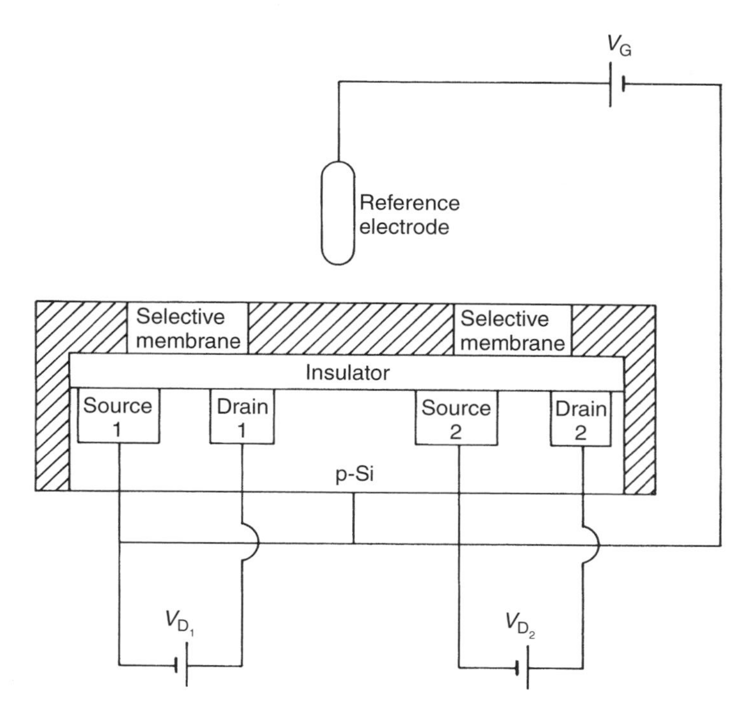

Figure 5.26 Schematic of a dual-gate ENFET, a semiconducting oxide sensor used to detect carbon monoxide and ethanol. From Eggins, B. R., *Biosensors: An Introduction*, Copyright 1996. © John Wiley & Sons Limited. Reproduced with permission.

The Pd–MOS FET device for hydrogen gas has been used with NAD^+–NADH and with hydrogen–hydrogenase linked to pyruvate–NH_3, which with alanine dehydrogenase gives alanine. This suggests a whole range of further possibilities for the many reactions which involve NADH.

SAQ 5.11

Explain the differences between IGFETs, CHEMFETs, ISFETs AND ENFETs.

Summary

This chapter presents the applications of four different types of electrochemically driven sensor. The first are those based on potentiometry, including ion-selective electrodes and potentiometric gas sensors. Under the class of amperometric sensors, voltammetric ion sensors and amperometric biosensors (including the role of mediators) are considered. First-, second- and third-generation biosensors are discussed, and amperometric gas sensors are included. In the third group, conducting sensors, including chemiresistors, biosensors and semiconducting oxide sensors are reviewed. Finally, the applications of field-effect transistors (FETs) are extended to direct chemical sensors (CHEMFETs), ion-selective sensors (ISFETs) and biosensors (ENFETs).

Further Reading

Albery, W. J. and Cranston, D. H., 'Amperometric enzyme electrodes', in *Biosensors: Fundamentals and Applications*, Turner, A. P. F., Karube, I. and Wilson, G. S. (Eds), Oxford University Press, Oxford, UK, 1987, pp. 180–210.

Bartlett, P. N., 'The use of electrochemical methods in the study of modified electrodes', in *Biosensors: Fundamentals and Applications*, Turner, A. P. F., Karube, I. and Wilson, G. S. (Eds), Oxford University Press, Oxford, UK, 1987, pp. 211–246.

Blackburn, G. F., 'Chemically sensitive field effect transistors', in *Biosensors: Fundamentals and Applications*, Turner, A. P. F., Karube, I. and Wilson, G. S. (Eds), Oxford University Press, Oxford, UK, 1987, pp. 481–530.

Bruckenstein, S. and Symanski, J. S., 'Analytical applications of gas membrane electrodes', *J. Chem. Soc., Faraday Trans. 1*, **82**, 1105–1116 (1986).

Cardosi, M. F. and Turner, A. P. F., 'The realisation of electron transfer from biological molecules to electrodes', in *Biosensors: Fundamentals and Applications*, Turner, A. P. F., Karube, I. and Wilson, G. S. (Eds), Oxford University Press, Oxford, UK, 1987, pp. 257–275.

Cass, A. E. G., Davis, G., Francis, G. D., Hill, H. A. O., Aston, W. J., Higgins, I. J., Plotkin, E. V., Scott, L. D. L. and Turner, A. P. F., 'Ferrocene-mediated enzyme electrode for amperometric determination of glucose', *Anal. Chem.*, **56**, 6567–6571 (1984).

Cassidy, J. F., Clinton, C., Breen, W., Foster, R. and O'Donohue, F., 'Novel electrochemical device for the determination of cholesterol or glucose', *Analyst,* **118**, 415–423 (1993).

Clark, L. C., 'The enzyme electrode', in *Biosensors: Fundamentals and Applications*, Turner, A. P. F., Karube, I. and Wilson, G. S. (Eds), Oxford University Press, Oxford, UK, 1987, pp. 3–12.

Davis, G., 'Cyclic voltammetry studies of enzymatic reactions for developing mediated biosensors', in *Biosensors: Fundamentals and Applications*, Turner, A. P. F., Karube, I. and Wilson, G. S. (Eds), Oxford University Press, Oxford, UK, 1987, pp. 247–256.

Enfors, S.-O., 'Compensated enzyme electrode for *in situ* process control', in *Biosensors: Fundamentals and Applications*, Turner, A. P. F., Karube, I. and Wilson, G. S. (Eds), Oxford University Press, Oxford, UK, 1987, pp. 347–355.

Hilditch, P. I. and Green, M. J., 'Disposable electrochemical biosensors', *Analyst*, **116**, 1217–1220 (1991).

Karube, I., 'Microbiosensors based on silicon technology fabrication', in *Biosensors: Fundamentals and Applications*, Turner, A. P. F., Karube, I. and Wilson, G. S. (Eds), Oxford University Press, Oxford, UK, 1987, pp. 471–480.

Kuan, S. S. and Guillbault, G. G., 'Ion selective electrodes and biosensors based on ISEs', in *Biosensors: Fundamentals and Applications*, Turner, A. P. F., Karube, I. and Wilson, G. S. (Eds), Oxford University Press, Oxford, UK, 1987, pp. 135–152.

Miasik, J. J., Hooper, A. and Tofield, B. C., 'Conducting polymer gas sensors', *J. Chem. Soc., Faraday Trans. 1*, **82**, 1117–1126 (1986).

Rusling, J. F. and Ito, K., 'Voltammetric determination of electron transfer rate between an enzyme and a mediator', *Anal. Chim. Acta,* **252**, 23–27 (1991).

Turner, A. P. F., Aston, W. J., Higgins, I. J., Bell, J. M., Colby, J., Davis, G. and Hill, H. A. O., 'Carbon monoxide: acceptor oxidoreductase from Pseudomonas thermocarboxydovorans strain C2 and its use in a carbon monoxide sensor', *Anal. Chim. Acta*, **163**, 161–174 (1984).

Wilson, R. and Turner, A. P. F., 'Glucose oxidase: an ideal enzyme', *Biosensors Bioelectron.*, **65**, 238–241 (1992).

Ye, L., Hammerle, M., Olsterhoorn, A. J. J., Schumann, W., Schmidt, H. -L., Duine, J. A. and Heller, A., 'High density "wired" quinoprotein glucose dehydrogenase electrode', *Anal. Chem.*, **65**, 238–241 (1993).

Chapter 6
Photometric Applications

Learning Objectives

- To understand the concept of 'evanescent' light.
- To distinguish between intrinsic and extrinsic methods.
- To appreciate the use of fluorescent and absorption methods in sensor applications.
- To understand the principles of solid-phase absorption sensors.
- To understand the operations of directly immobilized and indirectly immobilized reagents.
- To know how to apply photometric methods to sensors for pH, metal ions and gases (such as O_2, CO_2, SO_2, and NH_3).
- To understand how photometric methods can be used in biosensors.
- To appreciate the importance of reflectance methods.
- To know how the principles of reflectance are applied to attenuated total reflectance, total internal reflection fluorescence and surface plasmon resonance techniques.
- To acquire some knowledge of light scattering methods when applied to sensors.

6.1 Techniques for Optical Sensors

6.1.1 Modes of Operation of Waveguides in Sensors

Fibre optic waveguides for transmitting light to and from photometric sensors were introduced above in Chapter 2. These can operate in two modes, i.e. extrinsic and intrinsic.

6.1.1.1 The Extrinsic Mode

Here, the waveguide acts as both the light source and the collector. Modulation occurs externally from the fibre, as, e.g. in an in-line cell, such as that shown in Figure 6.1. The absorption of the light is described by the Beer–Lambert law, and thus is proportional to the measurand intensity and the thickness of the sample layer.

6.1.1.2 The Intrinsic Mode

In this mode, the measurand acts directly with the light in the waveguide. Phase, polarization and intensity may be modulated within the guide by a measurand that lies within the penetration depth for the *evanescent* field in the rarer medium adjacent to the guide, as shown in the model presented in Figure 6.2.

SAQ 6.1

Explain the difference between the 'intrinsic' and 'extrinsic' modes of operation of optical fibres.

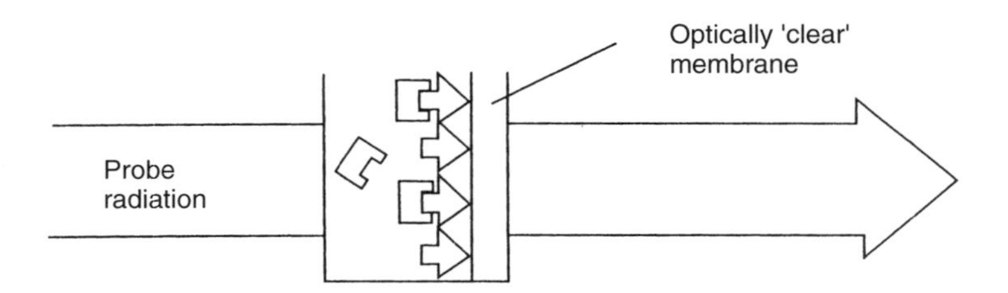

Figure 6.1 Schematic of an 'in-line' solid-state optical sensor cell. From Hall, E. A. H., *Biosensors*, Copyright 1990. © John Wiley & Sons Limited. Reproduced with permission.

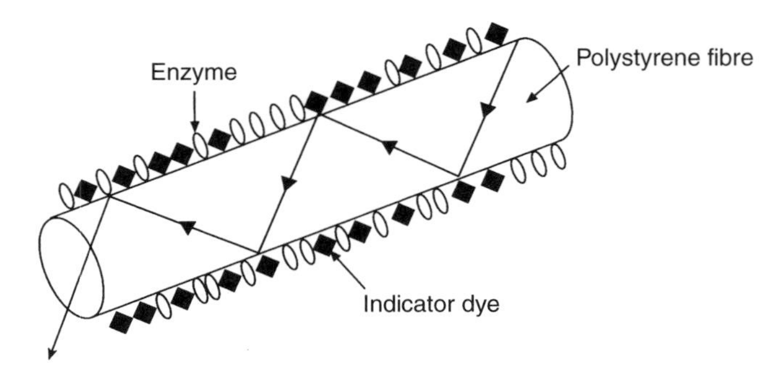

Figure 6.2 Schematic of a model for an intrinsic enzyme sensor based on a polystyrene fibre. From Hall, E. A. H., *Biosensors*, Copyright 1990. © John Wiley & Sons Limited. Reproduced with permission.

6.1.2 Immobilized Reagents

An immobilized reagent is normally required in optical sensors. This is often some type of dye which changes colour in the presence of the analyte. It can operate in a direct fashion, although sometimes indirect competitive binding has to be used. This is particularly useful if neither the analyte nor the reagent show a spectral change on binding, in which case an analyte analogue may be used.

6.1.2.1 Direct Method of Complexation of Analytes with Reagents

An analyte, A, can complex with a reagent, R, as follows:

$$\mathrm{A + R = AR}$$

where:

$$K = [\mathrm{AR}]/[\mathrm{A}][\mathrm{R}] \tag{6.1}$$

If {R} is the *total* reagent concentration, then:

$$\{\mathrm{R}\} = [\mathrm{R}] + [\mathrm{AR}]$$

and the free reagent concentration is given by the following:

$$[\mathrm{R}] = \{\mathrm{R}\}/(1 + K[\mathrm{A}]) \tag{6.2}$$

$$[\mathrm{AR}] = K[\mathrm{A}]\{\mathrm{R}\}/(1 + K[\mathrm{A}]) \tag{6.3}$$

There is no simple relationship between the signal due to the free reagent and the analyte, or the combined reagent plus analyte. If the measured optical parameter is proportional to [AR], then the response is proportional to [A] at low concentrations, such that $[\mathrm{A}] \ll 1/K$; if $[\mathrm{A}] \gg 1/K$, the response saturates to a limiting value. If the response is proportional to [R], the signal decreases with increasing [A]. A linear relationship in this case can be obtained from the following:

$$\{\mathrm{R}\}/[\mathrm{R}] = 1 + K[\mathrm{A}]$$

A better approach, therefore, is to use two wavelengths, i.e. one for AR and one for R (reference). Now we can write (from equation (6.1)):

$$[\mathrm{AR}]/[\mathrm{R}] = K[\mathrm{A}]$$

Hence, there is now a linear relationship between the ratio of the absorbances at λ_{AR} and λ_{R} and the concentration of A. This relationship is independent of [R].

DQ 6.1

Why is there no direct relationship between the analyte concentration and the absorbance when using an immobilized reagent?

Answer

The concentration of the analyte is a complicated function involving the free reagent and the complex formed between the reagent and the analyte. At low concentrations of A, the response due to the complex [AR] is proportional to the concentration of A (i.e. when $[A] \ll 1/K$ *in equation (6.3)). When* $[A] \gg 1/K$*, then the response reaches a limiting value proportional to [AR]/[R]. From equation (6.2), if the response is proportional to [R], the signal decreases with increasing [A]. A linear relationship would be* $\{R\}/[R] = 1 + K[A]$.

However if we measure the absorbancies at two wavelengths, i.e. one for AR and one for R, we can then use equation (6.1) and write [AR]/[R] = K[A], which now gives a linear relationship between the ratio of the absorbancies and the concentration of the analyte.

6.2 Visible Absorption Spectroscopy

6.2.1 Measurement of pH

The measurement of pH is fundamental to many applications. Very many dyes act as pH indicators. The greatest problem with such indicators is that the pH range of each of them is relatively small, i.e. less than 2 pH units. However, this does not matter too much with biochemical systems as they generally operate within fairly limited pH ranges. Methyl Red is a suitable (pH) dye, which has a distinctive visible spectrum with well-separated maxima for its acidic and basic forms (see Figure 6.3). One can see that both wavelengths are well up towards the maximum for the acid form of Methyl Red and well away from the basic form. Both reference wavelengths are in the region of negligible absorbancies for both species.

Methyl Red can be incorporated into polyacrylamide-coated microspheres, which in turn were used in the system illustrated in Figure 6.1. Detection was originally made at 558 nm, referenced to 600 nm, so giving a pH range of 7.0–7.5 ($\pm$0.01). This was to be used for measurement of the pH of whole blood whose pH ranges from 7.38 to 7.44. A later version used a light-emitting diode (LED) at 565 nm (bright green), referenced to an LED with $\lambda = 810$ nm (infrared). Detection was with a silicon p-i-n diode. Such a system made an inexpensive portable pH sensor.

SAQ 6.2

Are there any advantages of using photometric methods for pH measurement compared with potentiometric measurement?

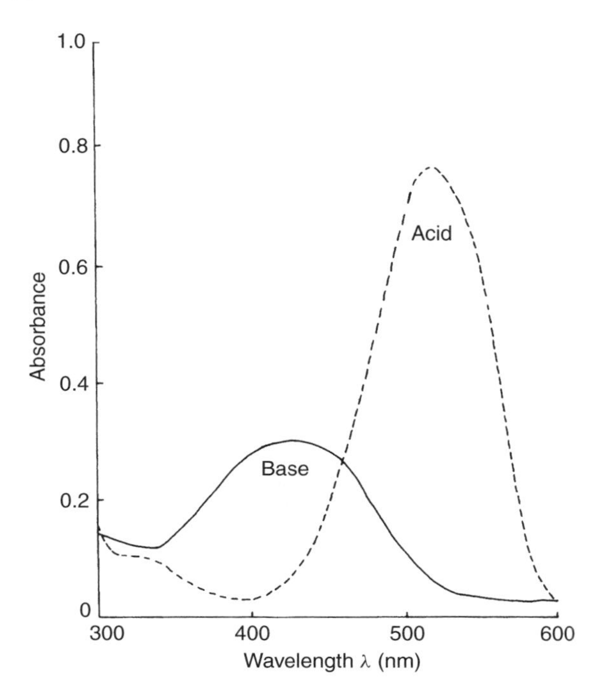

Figure 6.3 The ultraviolet and visible spectra of Methyl Red in its acid and base forms. From Hall, E. A. H., *Biosensors*, Copyright 1990. © John Wiley & Sons Limited. Reproduced with permission.

6.2.2 Measurement of Carbon Dioxide

Carbon dioxide can be determined by using a pH probe with a gas-permeable membrane containing a hydrogencarbonate buffer.

6.2.3 Measurement of Ammonia

A pH probe with a higher pH response is needed for ammonia, with 4-nitrophenol being a suitable indicator for such an application. A similar arrangement to that of the CO_2 sensor is used, except that in this case, of course, the buffer employed is ammonium chloride.

A different approach has been tried using oxazine perchlorate dye, which responds directly to NH_3. The dye is coated on a capillary waveguide and operates in the intrinsic mode (see Figure 6.4). Irradiation is with a 560 nm LED, and a photodiode detector is used.

The large range of NAD^+–NADH-related examples can also be included under this heading, as well as under the fluorescence mode.

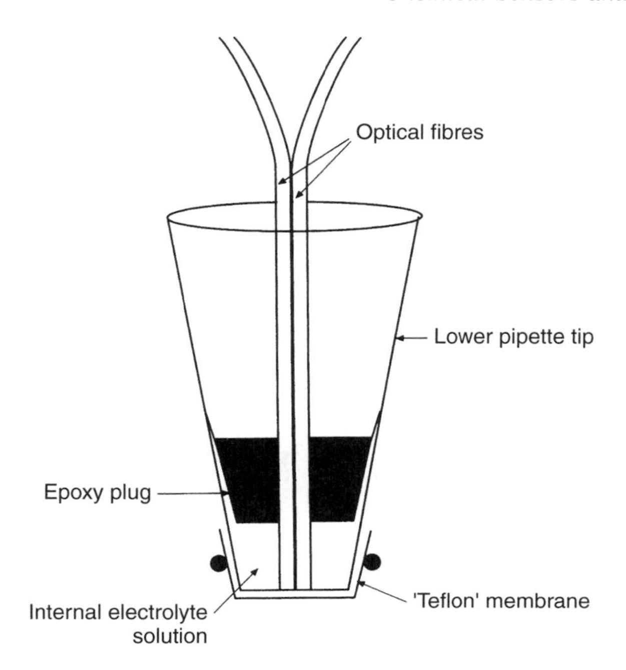

Figure 6.4 Schematic of a optrode, operating in the intrinsic mode, used for the determination of ammonia. From Hall, E. A. H., *Biosensors*, Copyright 1990. © John Wiley & Sons Limited. Reproduced with permission.

6.2.4 Examples That Have Been Used in Biosensors

A few of these systems have been developed into biosensors, e.g. for penicillin, urea and *p*-nitrophenyl phosphate.

6.2.4.1 Penicillin

Penicillinase catalyses the formation of penicilloic acid from penicillin with an increase in pH. A CNBr-activated optically clear cellulose membrane is covalently attached to the glutathione conjugate of Bromocresol Green. To this, penicillinase is bonded via a carbodiimide reaction.

6.3 Fluorescent Reagents

This is perhaps the most developed area so far for photometric sensors and biosensors.

6.3.1 Fluorescent Reagents for pH Measurements

Several fluorescent reagents have been developed for pH sensors, with perhaps the best of these being trisodium 8-hydroxy-1,3,6-trisulfonate. Absorption (excitation) bands occur at 405 (acid form) and 470 nm (base form), with emission (fluorescence) at 520 nm, as shown in Figure 6.5 All of these values are in the visible range, which enables less expensive optical components to be used. Such a system operates over the pH range 6.4–7.5.

6.3.2 Halides

A fluorescence-quenching sensor for Cl^-, Br^- and I^- determination has been designed, which uses either acridinium- or quinidinium-based fluorescent reagents covalently bound to a glass support via carbodiimide. The best sensitivity achieved was 0.15 mM for I^- when using the acridinium system.

6.3.3 Sodium

The optical label, fluoro(8-anilino-l-naphthalene sulfonate), forms ion-pairs with the immobilized ionophore–Na^+ complex, in competition with a quenching cationic polyelectrolyte, poly[copper(II)–polyethyleneimine], which suppresses the fluorescence of the excess label:

$$\underset{\text{(quenched)}}{\text{I–Na}^+ + \text{Poly-Fluor}_{\text{soln}}} \rightleftharpoons \underset{\text{(fluoresces)}}{\text{I–Na}^+\text{–Fluor}} + \text{Poly}_{\text{soln}}$$

where I is the ionophore.

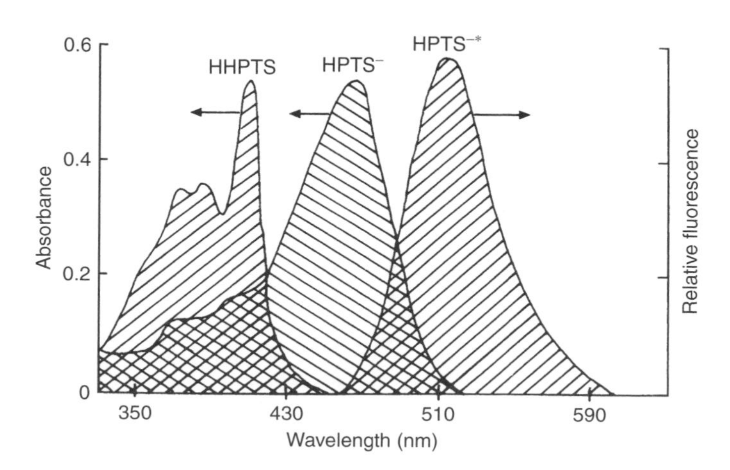

Figure 6.5 Absorption spectra of 8-hydroxy-1,3,6-trisulfonic acid (HPTS) in its acid (HHPTS) and base ($HPTS^-$) forms, and the fluorescence spectrum of the latter ($HPTS^{-*}$). From Hall, E. A. H., *Biosensors*, Copyright 1990. © John Wiley & Sons Limited. Reproduced with permission.

6.3.4 Potassium

A photo-activated crown compound, selective for K^+, has been made by reacting 2-hydroxy-1,3-xylyl-18-crown-5 with diazotised 4-nitroaniline, immobilized at the end of an optical fibre. Although the detection limit was 0.5 mM, the selectivity against Na^+ (=1/6) was insufficient for any practical clinical applications.

6.3.5 Gas Sensors

6.3.5.1 Oxygen

Oxygen is a very efficient quencher of fluorescence, so this property can be used in sensors in a number of different ways. One of the simplest involves perylenebutyric acid on a polyacrylamide support, which when excited at 468 nm fluoresces at 514 nm. This probe responds to oxygen over the pressure range 0–150 torr. An alternative material which is sometimes used is pyrenebutyric acid, but this requires excitation at 342 nm (in the UV region), which means that inexpensive plastic fibre optic components cannot be used. Another method, depending on the fluorescence *lifetime* rather than the *intensity*, has been developed for oxygen. This uses tris(2.2′-bipyridyl)ruthenium(II) dichloride hydrate, which produces relatively long-lived fluorescence at 610 nm when excited at 460 nm. The indicator was adsorbed on 'Kieselgel' in a silicone membrane. When using a blue LED source, the lifetime in the absence of oxygen was 205 ns. This sensor had an extended linear range and good long-term stability.

6.3.5.2 Sulfur Dioxide

Benzo[*b*]fluoranthene responds down to 84 ppm of SO_2 in the absence of oxygen (as found in exhaust gases), with the device arrangement being similar to that described above for oxygen.

SAQ 6.3

What are the advantages of fluorescent methods?

6.4 Indirect Methods Using Competitive Binding

Sometimes, a competitive binding method is used. This is particularly useful if neither the analyte nor the reagent show a spectral change on binding, in which case an *analyte analogue* may be used. This is a substance similar to the analyte which has inherent optical characteristics or which induces an optical change on bonding. The specific optical characteristic is often 'added' to the analogue as, for example, a fluorescent label. In the determination of glucose, a suitable 'reagent', often called a bio-receptor, is concanvalin A (Con A), while a suitable analogue is dextran labelled with fluorescein isothiocyanate (FITC–dextran). The Con A is immobilized on a cellulose hollow fibre (see Figure 6.6).

The FITC–dextran is freely mobile inside the optical cell, but cannot diffuse out through the membrane. The analyte can diffuse through the membrane into the cell because of the different relative sizes of the analyte and the analyte analogue (A*). In the absence of the sugar, the analyte analogue forms a complex with the receptor (R):

$$A^* + R \rightleftharpoons A^*\text{–}R$$

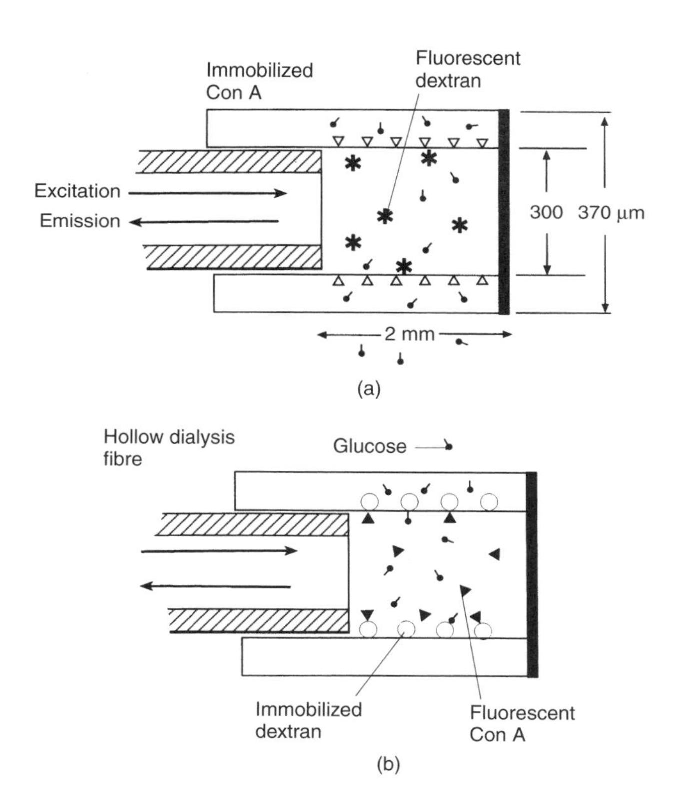

Figure 6.6 Two modes of operation of a biosensor for glucose which uses fluorescein isothiocyanate (FITC)-labelled dextran and Con A. From Schultz, J. S., Mansouri, S. and Goldstein, I. J., *Diabetes Care*, **5**, 245–253 (1982). Reproduced by permission of The American Diabetes Association, Inc.

The complex effectively removes some of the receptor from the walls of the cell so that a signal equivalent to 20% of the maximum fluorescence is observed. When the analyte (A) is introduced, there is a competing equilibrium between the analyte and the receptor:

$$A + R \rightleftharpoons A\text{–}R$$

This results in increasing amounts of receptor being displaced from the walls of the cell, thus resulting in further increases in the fluorescence signal. Eventually, all of the receptor is freed and no further increase in the fluorescence is seen – the signal is saturated. A typical response curve obtained for this system is shown in Figure 6.7.

We can analyse this behaviour as follows:

$$A + R \rightleftharpoons A\text{–}R$$

$$A^* + R \rightleftharpoons A^*\text{–}R$$

For simple unimolecular binding, the equilibrium constants for the two binding processes are given by the following:

$$K = [A\text{–}R]/[A][R]$$

$$K^* = [A^*\text{–}R]/[A^*][R]$$

If we let the total amount of the receptor be {R} and that of the analyte analogue be {A*}, these total concentrations are conserved, and so we can write the following:

$$\{R\} = [R] + [A\text{–}R] + [A^*\text{–}R]$$

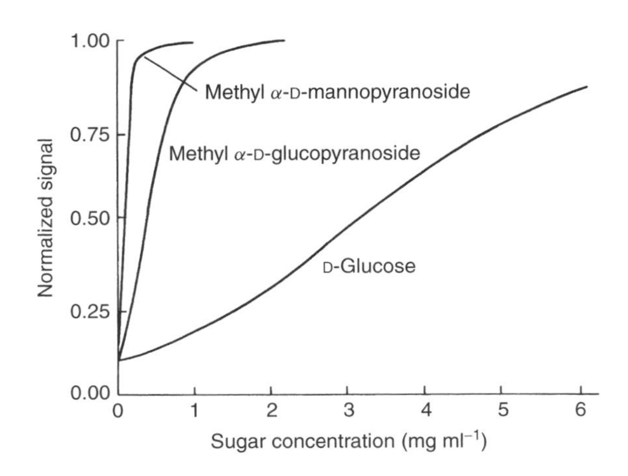

Figure 6.7 Response curves obtained for different sugars for a Con A-based biosensor. From Mansouri, S., *PhD Thesis*, University of Michigan, Ann Arbor, MI, USA, 1983.

and:

$$\{A^*\} = [A^*] + [A^*\text{–}R]$$

to express the mass balances of the two reagents. These four equations can be solved to find the ratio of the bound receptor to the total receptor concentrations, as follows:

$$([A^*]/\{A^*\})^2 + \{A^*\}/[A^*][(\{R\}/\{A^*\} - 1) + (K[A] + 1)/(\{A^*\}K^*)] - (K[A] + 1)/([A^*]K^*) = 0 \quad (6.4)$$

where $[A^*]/\{A^*\}$ represents the normalized response and has values between 0 and 1. The sensitivity is related to (A^*).

The normalized response is a function of two terms, i.e. $\{R\}/\{A^*\}$ and $(K[A] + 1)/(\{A^*\}K^*)$, where the concentration of the analyte, [A], occurs in the second term. Thus, a complete characterization of the model system can be obtained by a dimensionless plot of $[A]/\{A^*\}$ versus $(K[A] + 1)/(\{A^*\}K^*)$ for different values of $\{A^*\}/\{R^*\}$. Such a plot is shown in Figure 6.8.

The curves shown in this figure can be used to estimate the appropriate conditions for designing an optical sensor, by carrying out the following procedure:

(i) Estimate the mid-range analyte concentration of interest, $[A']$.

(ii) Select a bio-receptor with a binding constant of the order of $10/[A']$.

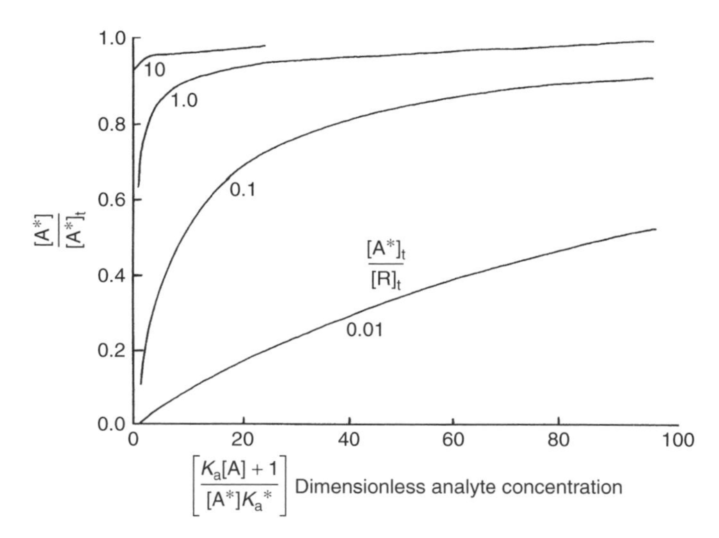

Figure 6.8 Parametric plots of equation (6.4): R, receptor; A, analyte; A^*, analyte analogue; K, rate constant; subscripts 't' and 'a' represent total and free analyte, respectively. Changes in $[A^*]_t/[R]_t$ show the effect of relative levels of analyte-analogue concentrations within the detector chamber (see text for further details). Reprinted from **Biosensors: Fundamentals and Applications** edited by A. P. F. Turner, I. Karube and G. S. Wilson (1987), by permission of Oxford University Press.

(iii) Estimate the minimum concentration of A* that can be estimated with the optical system, [A*′], and assume that {A*} is about 50[A*′].
(iv) Select (or synthesize) an analogue compound so that $K^* = 1/[A^{*\prime}]$.
(v) Develop a technique for loading the sensor compartment with the bio-receptor R such that the concentration of the sites is about 100{A*}.

Applying these principles to a glucose sensor, the range of blood glucose is 1–5 mg ml^{-1}, so [A′] is 2.5 mg ml^{-1} (0.025 M), while K for the Con A binding to glucose is 320 M^{-1}. The maximum amount of Con A that could be immobilized on the hollow surface of the dialysis fibre gave an effective concentration of 10^{-5} M. The binding constant K^* between FITC–dextran and Con A is about 7.5×10^{-4} M^{-1}, while the total FITC–dextran concentration {A*} is about 1.5×10^{-6} M. The value of $K[A']$ was ca. 6, with glucose at a level of 2.5 mg ml^{-1}, and {R}/{A*} was 7. These data fit the above criteria.

6.5 Reflectance Methods – Internal Reflectance Spectroscopy

6.5.1 Evanescent Waves

In order to reflect light totally within an optical fibre, there needs to be a zero flux of energy into the 'optically rarer' medium (the cladding). In fact, there is a finite decaying electrical field across the interface. Thus, a fraction of the radiation extends a short distance from the guiding region into the medium of lower refractive index that surrounds it. The evanescent wave (EW) field decays exponentially with distance from the waveguide interface. This defines a short-range sensing volume within which the evanescent energy may interact with the molecular species that are present.

DQ 6.2

What are the advantages of 'evanescent light' sensors?

Answer

These can be summarized as follows:

(i) No coupling optics are needed.
(ii) Considerable miniaturization is possible.
(iii) Potential greater control of launch optics. This allows one to confine the evanescent field to a short distance from the surface, so that discrimination between surface and bulk effects is possible.
(iv) Enhanced sensitivity over conventional optical methods because of the greater power available in the evanescent field.

(v) *The very small pathlength is less affected by scattering.*
(vi) *Greater design control is possible.*

6.5.2 Reflectance Methods

Such methods are concerned with studying material adsorbed on an optical surface, and are particularly suitable for use in immunoassays. There are three principle variations, as follows:

- Attenuated total reflectance (ATR)
- Total internal reflection fluorescence (TIRF)
- Surface plasmon resonance (SPR)

The basic arrangement for attenuated total reflectance is shown in Figure 6.9 (also see Figure 2.44 above).

With a light wave striking the interface between two media with refractive indices of n_1 and n_2, total internal reflection occurs when the angle of reflection θ is such that:

$$\sin \theta_c = n_2/n_1, \text{ and } n_1 > n_2$$

If $\theta > \theta_c$, there is an evanescent wave refracted through the interface in the z-direction which penetrates the n_2 medium a distance d_p, which is of the order of a wavelength. It can be shown that the electrical field vector of this wave (E) is largest at the interface (E_0) and decays exponentially with distance (z), as shown in the following equation (Figure 6.10):

$$E = E_0 \exp(z/d_p) \tag{6.5}$$

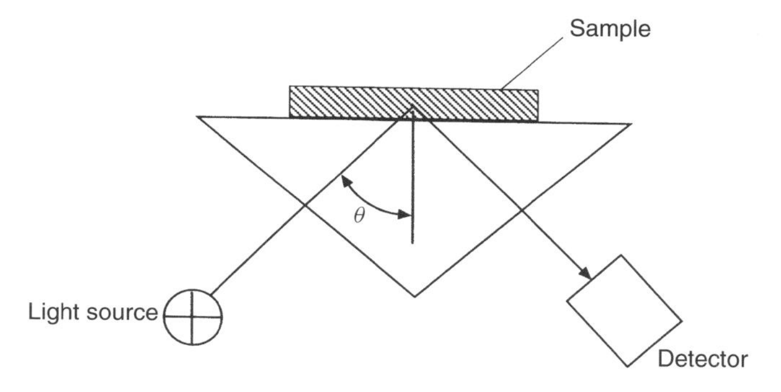

Figure 6.9 Basic arrangement for carrying out attenuated total reflectance measurements. Reprinted from **Biosensors: Fundamentals and Applications** edited by A. P. F. Turner, I. Karube and G. S. Wilson (1987), by permission of Oxford University Press.

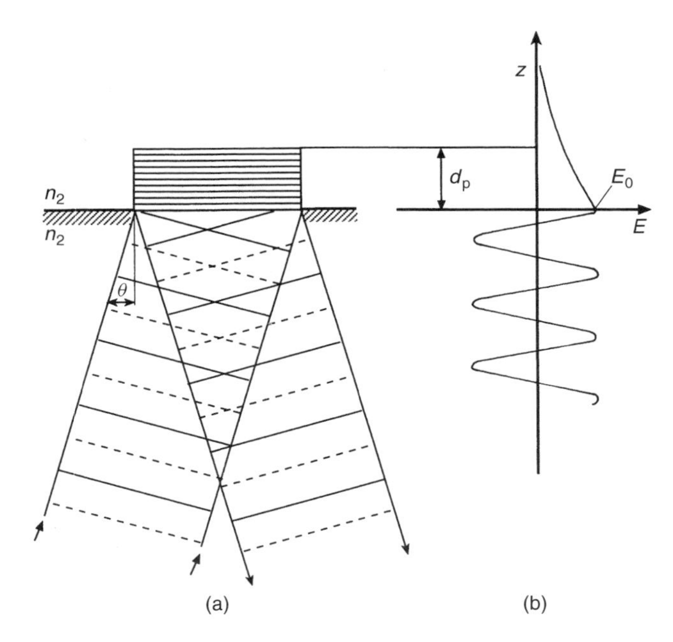

Figure 6.10 Generation of the evanescent wave at an interface between two optical media. Reprinted from **Biosensors: Fundamentals and Applications** edited by A. P. F. Turner, I. Karube and G. S. Wilson (1987), by permission of Oxford University Press.

The depth of penetration, d_p, can be related to other factors by the following:

$$d_p = \frac{\lambda/n_1}{2\pi[\sin^2\theta - (n_2/n_1)^2]^{1/2}} \tag{6.6}$$

It can be seen from equation (6.6) that, qualitatively, d_p decreases with increasing θ and increases as n_2/n_1 tends to unity. We can therefore select a value of d_p by an appropriate choice of values of θ, n_1 and λ. For example, with a quartz waveguide, $n_1 = 1.46$ and, if the sample is in water, $n_2 = 1.34$. This gives a value for θ_c of 66°. If θ is set to 70° and λ to 500 nm, the value of d_p is 270 nm, which will easily contain a monolayer of an immunological component with a diameter of about 25 nm.

However, d_p is just one of the four factors determining the attenuation of reflection. The others are the polarization-dependent electric field intensity at the reflecting surface, the sampling area and the matching of the two refractive indices, n_1 and n_2. An effective thickness of d_e takes account of all these factors. This represents the thickness of the film required to produce the same absorption in a transmission experiment.

In order to enhance the sensitivity, multiple reflections can be employed, as shown in Figure 6.11. If the number of reflections (N) is a function of the length

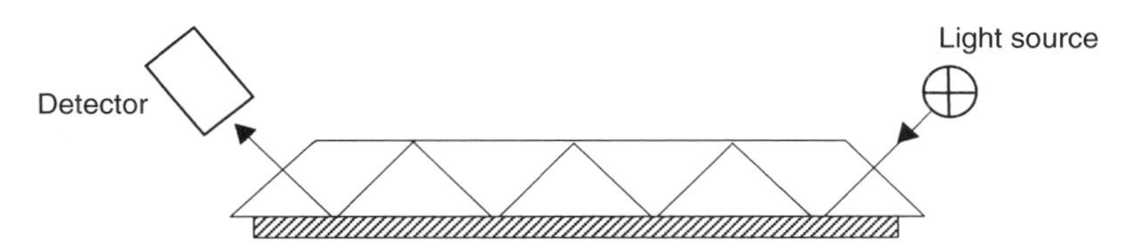

Figure 6.11 Total internal reflection showing multiple reflections. Reprinted from **Biosensors: Fundamentals and Applications** edited by A. P. F. Turner, I. Karube and G. S. Wilson (1987), by permission of Oxford University Press.

(l) and thickness (T) of the waveguide, and the angle of incidence (θ), so that the following applies:

$$N = (l/T)\cot\theta$$

then the longer and thinner the wave guide, the larger is N and the more frequently the evanescent wave interacts with the surface layer of analyte. If R is the reflectivity, then we can write:

$$R = 1 - \alpha d_e$$

where α is the absorption coefficient. Thus, for N reflections, we have:

$$R^N = 1 - N\alpha d_e$$

SAQ 6.4

What is 'evanescent light'?

Figures 6.12 and 6.13 show two practical arrangements used for internal reflection elements (IREs). The first of these involves multiple internal reflections, while the second system uses a fibre optic assembly. Both of these systems could be used for ATR or SPR techniques.

6.5.3 Attenuated Total Reflectance

Here, an absorbing material is placed in contact with the reflecting surface of an internal reflection element (IRE), thus causing attenuation of the internally reflected light. The intensity of the light is measured against that of the incident wavelength. Attenuated total reflection (ATR) has been used for monitoring immunoassays in the IR, visible and UV regions. For example, haemoglobin and rabbit antihaemoglobin antisera have been monitored by using this approach. The rabbit antibodies were covalently immobilized on the surface of a quartz microscope slide and then reacted with different concentrations of haemoglobin. The

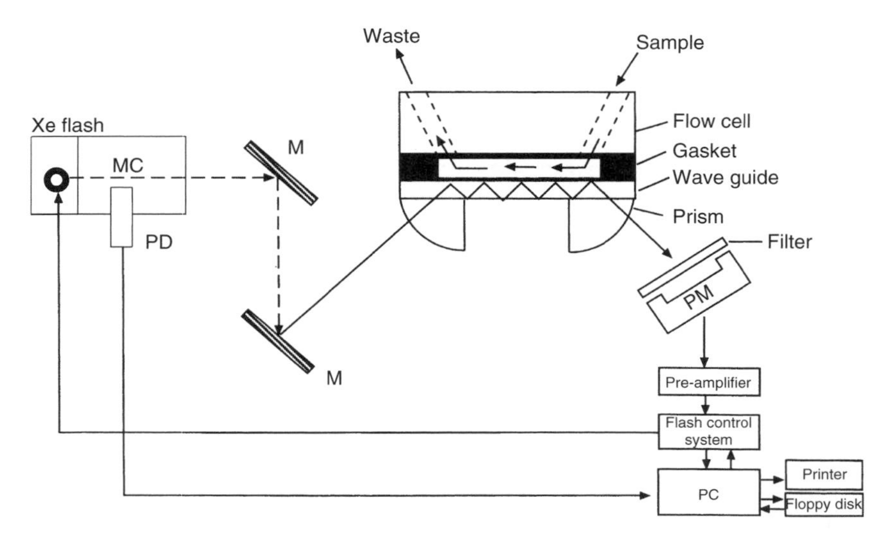

Figure 6.12 Schematic of the instrumental layout used for measuring immunoassays with a multiple internal reflection plate: PM, photomultiplier tube; PD, photodiode; MC, monochromator; M, mirror. Reprinted from **Biosensors: Fundamentals and Applications** edited by A. P. F. Turner, I. Karube and G. S. Wilson (1987), by permission of Oxford University Press.

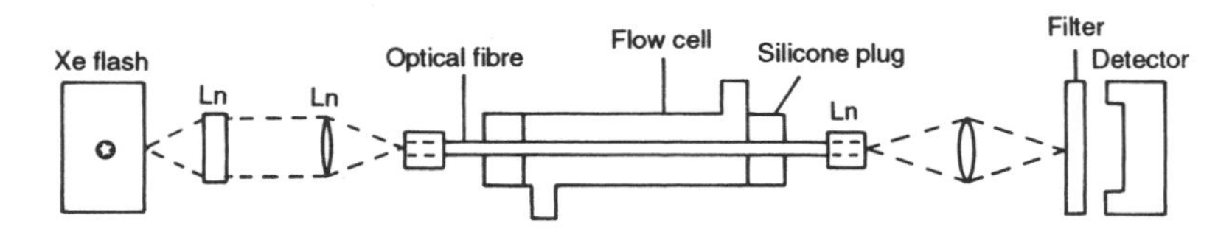

Figure 6.13 Schematic of a fibre optic assembly employing a flow cell and light coupling optics used for measuring immunoassays; Ln, lens. Reprinted from **Biosensors: Fundamentals and Applications** edited by A. P. F. Turner, I. Karube and G. S. Wilson (1987), by permission of Oxford University Press.

attenuation of the reflected light was measured at 410 nm. A similar application worked at 310 nm.

6.5.4 Total Internal Reflection Fluorescence

In this technique, the emitted fluorescence can be detected either by a detector positioned at right angles to the interface (Figure 6.14(a)), or in line with the primary beam (Figure 6.14(b)). In fact, it can be shown, both theoretically and experimentally, that the latter method results in an enhancement of up to 50-fold in the emission signal. This is particularly useful when using an optical fibre as the IRE. In addition, by avoiding measurement of the fluorescence through the bulk

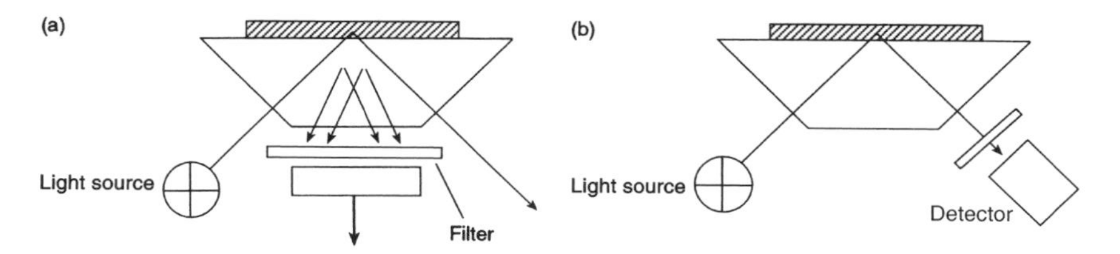

Figure 6.14 Schematics of the arrangements used for the detection of internal reflection by using (a) right-angled and (b) in-line fluorescence. Reprinted from **Biosensors: Fundamentals and Applications** edited by A. P. F. Turner, I. Karube and G. S. Wilson (1987), by permission of Oxford University Press.

of the sample solution around the fibre, interference is minimized. Total internal reflection fluorescence (TIRF) has also been used to measure immunological reactions. For example, phenylarsonic acid was immobilized on the surface of a quartz microscope slide via a hapten–albumin conjugate. The FITC-labelled antibody, which could bind to the immobilized hapten, could be detected by exciting fluorescence at the surface with the evanescent wave.

6.5.5 Surface Plasmon Resonance

6.5.5.1 Brewster Angle Measurements

For plane polarized light, at the Brewster angle, which is defined as $\theta = \tan^{-1}(n_1/n_2)$, the reflectance $(R_p) = 0$. However, for a transparent incident phase and an adsorbing substrate, n_2 is given by $N_2 = n_2 + i/k_2$ $(i = (-1)^{1/2};\ k_2 = 0)$ and R_p shows a minimum at the *pseudo-Brewster* angle. The angle for which R_p is a minimum $(R_{p,min})$ is very sensitive to overlayers on the surface, so we can measure the change in $R(=\Delta R)$, as follows:

$$\Delta d_{min} = (\Delta R/\Delta d)^{-1}\Delta R_{min}$$

where d is the thickness of the surface layer. If ΔR_{min} is ca. $\pm 0.025\%$, then d_{min} is ca. 0.05 nm. If this is used for an antibody–antigen reaction with concentrations of about 0.8 $\mu g\ cm^{-2}$, it has been suggested that the detection limit would be about 0.02–0.05 $\mu g\ cm^{-2}$, corresponding to a thickness change of 0.2–0.4 nm.

6.5.5.2 Surface Plasmons

These are formed in the boundary of a solid (metal or semiconductor), where the electrons behave like a *quasi-free* electron gas. Exterior electrical fields in the boundary produce quanta of oscillations of surface charges, and such charge oscillations couple with high-frequency electromagnetic fields extending into space.

These plasmons can be excited by light or by electron beams. The most useful are non-radiative plasmons excited by evanescent light waves. The type of plasmon is characterized by an exponential decrease in the electric field with increasing distance from the boundary.

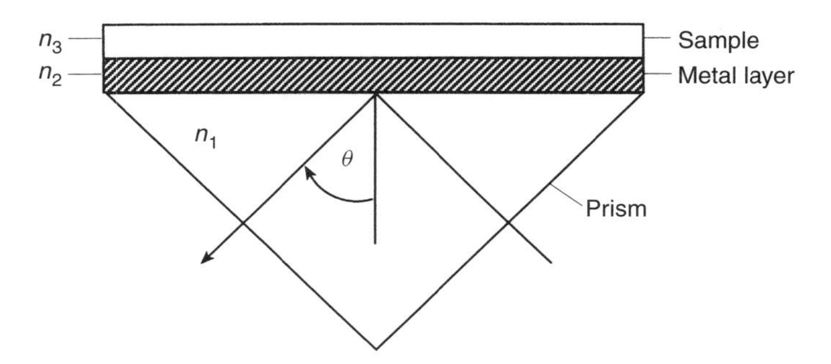

Figure 6.15 Schematic of the attenuated total reflection method of exciting non-radiative plasmons using the *Kretschmann* arrangement: n_1, n_2 and n_3 are the refractive indices of the glass prism, metal layer and sample, respectively. From Kretschmann, E. and Raether, H., *Naturforschung*, **122**, 2135–2136 (1968). Reproduced by permission of Verlag der Zeitschrift für Naturforschung.

The normal SPR technique uses the basic ATR configuration. In this, a prism of refractive index n_1 is coated with a very thin layer of metal, such as 60 nm of silver, with a refractive index n_2, on to which is deposited a layer of sample of refractive index n_3, so that $n_1 > n_3$ (as shown in Figure 6.15).

The plane-polarized incident field has an angle θ such that the photon momentum along the surface matches the plasmon frequency. Thus, the light can couple to the electron plasma in the metal. This is known as *surface plasmon resonance*.

SAQ 6.5

What is meant by the Brewster angle?

SAQ 6.6

What is a surface plasmon?

The intensity of the totally reflected light is measured. This shows a sharp drop with increase in θ, depending on the depth and width and the characteristic absorbance and thickness of the metal. Such an effect can be seen in Figure 6.16, which illustrates the distribution of energy density $|H(x)|^2$ across the metal layer for different angles of incidence.

For $\theta_0 = 50^\circ$, the energy is outside the resonance region and it decays exponentially inside the plasma. For $\theta_0 = 45.4^\circ$, it is near to the resonance and the energy dips to a minimum and then rises to a higher value at the boundary. When $\theta_0 = 45.20$, resonance is achieved, so the field energy rises to a maximum

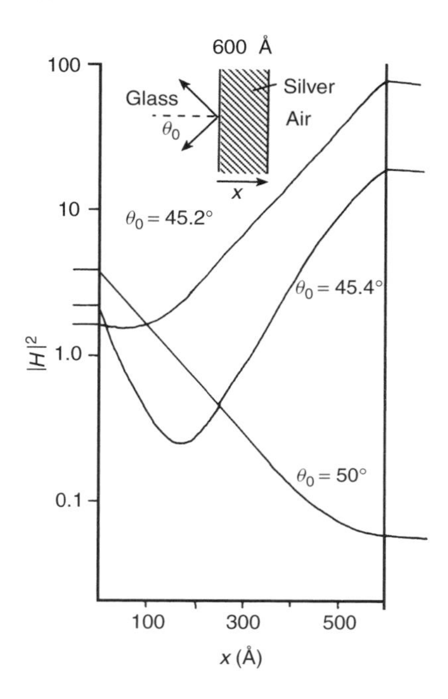

Figure 6.16 The calculated electromagnetic density $|H(x)|^2$ in a 600 Å silver film as a function of the angle of incidence θ_0. From Raether, H., 'Excitation of plasmons and interband transitions by electrons', in *Springer Tracts in Modern Physics*, Volume 88, 1980. © Springer-Verlag GmbH & Co. KG. Reproduced with permission.

at the boundary, with a value of about 80 times the value without resonance. Figure 6.17 shows plots of intensity (R_p) versus the angle of incidence (θ).

It can be seen that as the adsorbed layer becomes thicker, the angle for resonance shifts progressively to larger values and the width of each peak increases. If the optical system can resolve better than 0.05°, coatings of a few angstroms in thickness can be measured. The angular position is also very sensitive to changes in refractive index just outside of the metal film. Thus, changing from air ($n = 1.0$) to water ($n = 1.33$) causes a shift in resonance angle of 25°.

The experimental arrangement for SPR is similar to that used for ATR (as shown in Figure 6.15 above).

The IRE prisms, which are usually made of quartz, are coated with a thin layer of a metal such as Au, Ag, Al or Cu. One can achieve the point of resonance in several ways. First, we can keep the wavelength constant and vary the angle to obtain maximum intensity of the reflected wave. Angular shifts as small as 0.0005° have been detected in this way. An alternative way is to keep the angle constant and vary the wavelength to obtain the same effect. Light scattered by

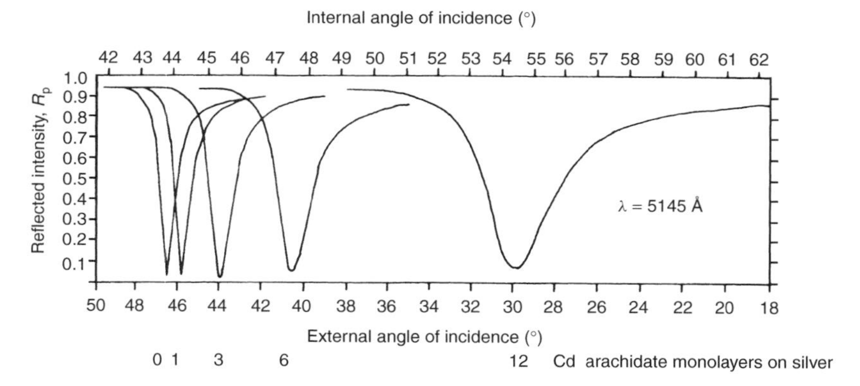

Figure 6.17 Attenuated total reflectivity curves obtained for silver films covered with different amounts of arachidate monolayers. Reprinted from *Surf. Sci.*, **74**, Pockrand, I., Swalen, J. D., Gordon II, J. G. and Philpott, M. R., 'Surface plasmon spectroscopy of organic monolayer assemblies', 237–244, Copyright (1977), with permission from Elsevier Science.

the surface roughness of the film may be measured against the incident angle at fixed wavelength, using the same geometry as that shown above in Figure 6.15.

The main applications of SPR so far have been in immunoassays. One of the immunological pair has to be immobilized to the IRE surface, which can be a difficult procedure. The most commonly used methods are adsorption and covalent bonding (see Chapter 3 earlier).

A model example is the reaction between human immunoglobulin (class G) (IgG) and anti-IgG. Here, the antigen is adsorbed on to the silvered surface of a prism to give a protein layer which is ca. 50 Å thick. Various concentrations of anti-IgG were incubated with the adsorbed layer and the shift in resonance angle measured at a fixed incident angle, as shown in Figure 6.18.

Surface plasmon resonance is a relatively new technique that has great potential for sensor applications. An SPR device, known as the 'BIAcore' is marketed by Pharmacia Biosensors (Sollentuna, Sweden).

SAQ 6.7

What are the advantages of surface plasmon resonance?

DQ 6.3

What are the factors limiting the performance of optical chemical sensing?

Answer

(i) They only work if the appropriate reagent can be developed.

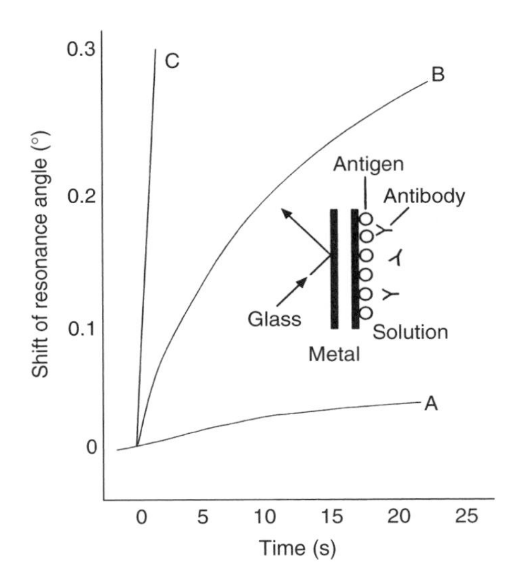

Figure 6.18 Shift in resonance as a function of time for three anti-IgG concentrations: A, 2 μg ml^{-1}; B, 20 μg ml^{-1}; 200 μg ml^{1}. The inset illustrates the antibody-binding event in the reaction between human IgG and anti-IgG. Reprinted from *Sensors Actuators*, **4**, Liedberg, B., Nylander, C. and Lundström, I., 'Surface plasmon resonance for gas detection and biosensing', 299–304, Copyright (1983), with permission from Elsevier Science.

(ii) *They are subject to ambient background light interference.*
(iii) *They have a limited dynamic range – typically 10^2 compared with 10^6 to 10^{12} for ISEs.*
(iv) *They are extensive devices, and dependent on the amount of reagent, and hence they may be difficult to miniaturize.*
(v) *Response times may be slow because of the time of mass transfer of analytes to the reagent phase.*
(vi) *Problems with long-term calibration of the sensor due to such factors as:*
- *Stability of reagents under incident light;*
- *Leaching of reagents;*
- *Stability of immobilization matrix;*
- *Susceptibility to interference and fouling.*

6.6 Light Scattering Techniques

6.6.1 Types of Light Scattering

The application of laser light has revolutionized the use of light scattering techniques. There are a number of modes of scattering, such as Rayleigh scattering,

Rayleigh–Gans–Debye scattering, and Mie scattering. Many analytical techniques make use of light scattering. There are two main forms, i.e. *static* and *dynamic* light scattering, where the latter is more useful for applications involving biological materials. There are three relatively new and powerful techniques in this category, namely:

- Quasi-elastic light scattering spectroscopy (QELS)
- Photon correlation spectroscopy (PCS)
- Laser Doppler velocimetry (LDV)

These will be discussed in the following sections.

6.6.2 Quasi-Elastic Light Scattering Spectroscopy

An assembly of particles suspended in a medium move under Brownian motion and the different ways in which light is scattered from each particle cause a fluctuation in the intensity with time. This distribution contains information about the size and size distribution of the particles. The time-scale of the intensity fluctuations is the basis of dynamic light scattering or *intensity fluctuation spectroscopy*, known as *quasi-elastic light scattering spectroscopy* (QELS). This has been used for the particle size determination of micelles and charge proteins.

6.6.3 Photon Correlation Spectroscopy

This technique is really an extension of QELS, with the experimental set-up used being shown in Figure 6.19. The continuous laser light is passed through a cell

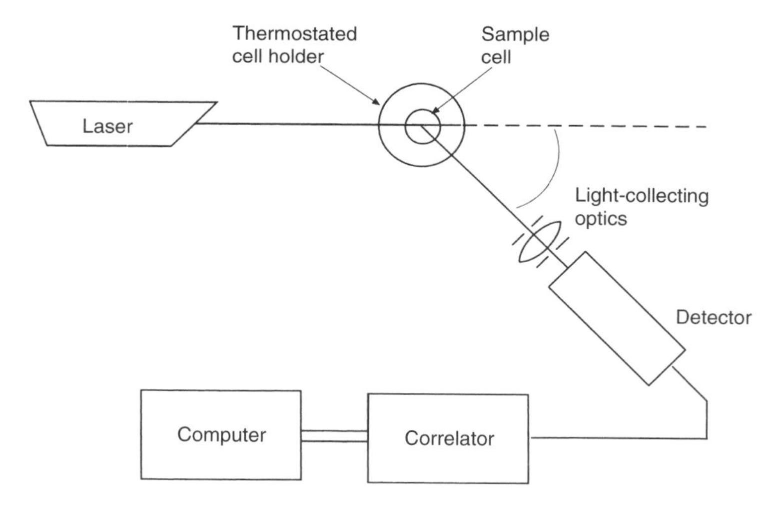

Figure 6.19 Schematic of the instrumental layout used to carry out photon correlation spectroscopy. Reprinted from **Biosensors: Fundamentals and Applications** edited by A. P. F. Turner, I. Karube and G. S. Wilson (1987), by permission of Oxford University Press.

containing the sample, which will be in suspension under Brownian motion. The scattered laser light is collected by a lens and converted by a photo-detector into an electrical signal. This signal is then analysed by a photon-correlator. The latter computes averages of the signal compared with itself at different delay times. This is called the *autocorrelation function* of the signal. This usually decays exponentially with the delay time, which is related to the size of the scattered particles or macromolecules. From this function, a particle size distribution is constructed. Photon correlation spectroscopy (PCS) operates with between 10^4 and 10^{10} particles/ml, and particle sizes between nanometres and micrometres.

For immunoassays, PCS offers up to 100–1000-fold increases in sensitivity compared with conventional methods, and is comparable in sensitivity to radioimmunoassays, but without involving radioisotopes. There are many other applications to biomacromolecules, particularly with regard to the determination of the sizes and shapes of the particles, which can be of assistance in the determination of the tertiary structure and behaviour of biological macromolecules.

6.6.4 Laser Doppler Velocimetry

This technique is used to obtain information about the velocity of particles flowing, for example, through a tube. The arrangement for this is shown in Figure 6.20.

Continuous plane polarized laser light (a few mW in power) is split into two equal beams, which are then focused to intersect in the fluid flow. The particles in the flow scatter the light from each laser beam with slightly different Doppler frequencies. A detector, producing beats at the photo-detector, collects some of the scattered light. Analysis of the frequency data received can give an estimate of the velocities of the particles. Laser Doppler velocimetry (LDV) has been applied

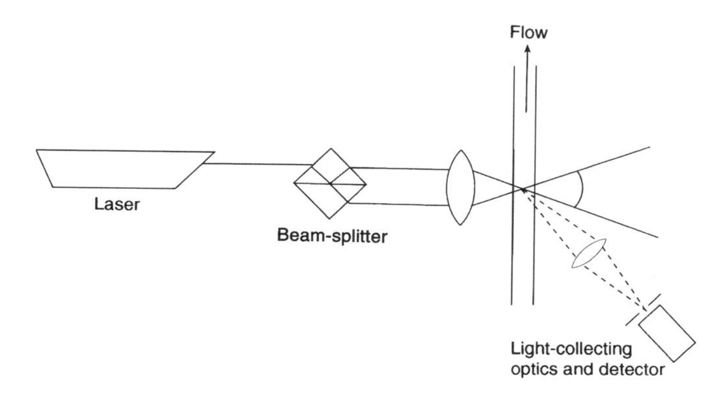

Figure 6.20 Schematic of the instrumental layout used to carry out laser Doppler velocimetry. Reprinted from **Biosensors: Fundamentals and Applications** edited by A. P. F. Turner, I. Karube and G. S. Wilson (1987), by permission of Oxford University Press.

to electophoretic light-scattering studies of living cells, vesicles and counter-ion condensation on to DNA.

At present, all the above techniques require large and bulky components and need frequency recalibration (except PCS), a high level of operator expertise and considerable sample preparation. However, with the application of optical fibres, it should be possible to simplify the procedures and hopefully permit the development of biosensors making use of these methods.

SAQ 6.8

What are the main types of light scattering?

Summary

This chapter discusses advanced applications of photometric techniques to sensors. These involve the use of optical-fibre waveguides, which can be used in either intrinsic or extrinsic modes, introducing the concept of 'evanescent' light. Solid-phase absorption methods are presented, including the use of directly and indirectly immobilized reagents. Fluorescent and absorption methods are compared. The application of these methods to sensors for pH, metal ions and gases such as O_2, CO_2, SO_2, and NH_3, is described. Techniques for using photometric methods in biosensors are also presented.

The principles of reflectance methods are described, and applied to attenuated total reflectance (ATR), total internal reflection fluorescence (TIRF) and surface plasmon resonance (SPR). Some applications of these techniques are given. In addition, an outline of light scattering methods for sensors are presented, including quasi-elastic light scattering spectroscopy, photon correlation spectroscopy and laser Doppler velocimetry.

Further Reading

Blum, L. J., Gautier, S. M. and Coulet, P. R., 'Fiber-optic biosensors based on luminometric detection', *Food Sci. Technol.*, **60**, 101–121 (1994).

Carr, R. J. G., Brown, R. O. W., Rarity, I. G. and Clarke, D. J., 'Laser light scattering and related techniques', in *Biosensors: Fundamentals and Applications*, Turner, A. P. F., Karube, I. and Wilson, G. S. (Eds), Oxford University Press, Oxford, UK, 1987, pp. 679–703.

Lavers, C. R., Itoh, K., Wu, S. C., Murabayashi, M., Mauchline, I., Stewart, D. and Stout, T., 'Planar optical waveguides for sensing applications', *Sensors Actuators, B*, **69**, 85–95 (2000).

McCapra, E., 'Potential applications of bioluminescence and chemiluminescence in biosensors', in *Biosensors: Fundamentals and Applications*, Turner, A. P. F., Karube, I. and Wilson, G. S. (Eds), Oxford University Press, Oxford, UK, 1987, pp. 617–638.

McCraith, B. D., 'Optical chemical sensors', in *Chemical and Biological Sensors*, Diamond, D. (Ed.), Wiley, New York, 1998, pp. 195–233.

Schultz, J. S., 'Design of fibre optic biosensors based on bioreceptors', in *Biosensors: Fundamentals and Applications*, Turner, A. P. F., Karube, I. and Wilson, G. S. (Eds), Oxford University Press, Oxford, UK, 1987, pp. 638–655.

Seitz, W. R., 'Optical sensors based on immobilised reagents', in *Biosensors: Fundamentals and Applications*, Turner, A. P. F., Karube, I. and Wilson, G. S. (Eds), Oxford University Press, Oxford, UK, 1987, pp. 599–617.

Sutherland, R. M. and Dahne, C., 'IRS devices for optical immunoassays', in *Biosensors: Fundamentals and Applications*, Turner, A. P. F., Karube, I. and Wilson, G. S. (Eds), Oxford University Press, Oxford, UK, 1987, pp. 655–679.

Chapter 7

Mass-Sensitive and Thermal Sensors

Learning Objectives

- To be able to explain the principles of the piezo-electric effect.
- To be able to explain how the variation in frequency with mass can be used in sensors.
- To know which materials can be used to produce the piezo-electric effect.
- To know how the piezo-electric effect can be applied in gas analysis and in biosensors.
- To be able to explain the application of piezo-electricity to the quartz crystal microbalance and the electrochemical quartz crystal microbalance.
- To describe the modification of the piezo-electric effect in the formation of surface acoustic waves.
- To understand the variations on the surface acoustic waves principle, i.e. plate wave mode, evanescent mode, Lamb mode and shear mode.
- To know how these variations can be applied to vapour analysis and in biosensors.
- To know the three main types of thermal sensing devices, i.e. calorimetric, catalytic and thermal conductivity.
- To know the construction and operation of a thermistor.
- To describe the application of the thermistor in biosensors.
- To know the principles and applications of the catalytic gas sensor.
- To describe the pellister.
- To describe thermal conductivity devices as used in gas chromatography.

7.1 The Piezo-Electric Effect

7.1.1 Principles

In 1880, The Curie brothers discovered that anisotropic crystals, i.e. those with no centre of symmetry, such as quartz and tourmaline, give out an electrical signal when mechanically stressed. Conversely, if an electrical signal is applied to such crystals, they will deform mechanically. Thus, with the application of an oscillating electrical potential, the crystal will vibrate.

Every crystal has its own natural resonant frequency of oscillation, which can be modulated by its environment. The usual value of this frequency is in the 10 MHz region, i.e. radiofrequency. The actual frequency is dependent on the mass of the crystal, together with any other material coated on it. The change in resonant frequency (Δf) resulting from the adsorption of an analyte on its surface can be measured with high sensitivity (500–2500 Hz g^{-1}), and when applied in sensors can thus result in devices with pg detection limits.

The relationship between the surface mass change, Δm, and the change in resonant frequency, Δf, is given by the Sauerbrey equation, as follows:

$$\Delta f = -2.3 \times 10^6 f^2 \Delta m/A$$

where Δm is the mass in grams of the adsorbed material on an area A (cm^2) of the sensing region, and f is the overall resonant frequency. For a 15 kHz crystal, a resolution of 2500 Hz μg^{-1} is likely, so that a detection limit of 10^{-12} g (1 pg) is achievable.

Materials that show the piezo-electric effect and can be used in sensor devices now include ceramic materials such as barium and lead titanates, as well as the 'natural' materials mentioned above. Some organic polymers, such as poly (vinylidene fluoride) (PVDF) $(-CF_2-CH_2-CF_2-)_n$, also form crystals with piezo-electric properties.

A schematic of a typical arrangement used in a piezo-electric sensor device is shown in Figure 7.1.

SAQ 7.1

(a) What is meant by the piezo-electric effect?
(b) Name three materials that display piezo-electric properties.

7.1.2 Gas Sensor Applications

Such sensors are particularly useful for the analysis of vapours and gases that can be adsorbed on to a surface coating on the crystal. Although this method can display a lack of selectivity, specific coatings, however, may selectively adsorb certain gases. Thus, hygroscopic coatings, such as gelatine, silica gel, and

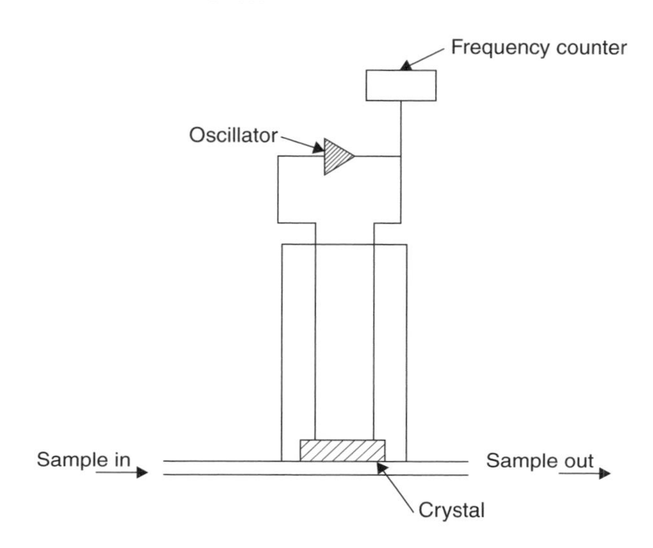

Figure 7.1 Schematic of a typical arrangement used in a piezo-electric sensor device.

molecular sieves, will adsorb water vapour. These substances must be reversibly removable so that the sensor can be used repeatedly. Such sensors have been successfully developed commercially for use as devices for water vapour detection.

Sulfur dioxide is a major pollutant in air, resulting from the combustion of fossil fuels and from motor vehicle engine exhausts. Organic amines will adsorb sulfur dioxide in a reversible manner, according to the following:

$$R_3N + SO_2 + H_2O \rightleftharpoons R_3NH^+HSO_3^-$$

However, such amines will also adsorb other acid gases, such as the oxides of nitrogen.

Highly selective sensors have been made for hydrogen sulfide, a very toxic gas with a foul odour, even at very low concentrations. This gas is still toxic at levels that are not detectable by the human nose. Metal acetates, such as those of copper, silver or lead, have been successfully used to detect hydrogen sulfide in sensor devices.

Carbon monoxide is a very dangerous toxic gas, as it has no detectable odour – this gas can often be produced by faulty heating appliances. Although it is not very reactive, its reducing properties have been used to make a piezo-electric sensor. The gas reacts with mercury(II) oxide at 210°C, producing mercury vapour, as follows:

$$HgO + CO = Hg + CO_2$$

The mercury is then sensed by using a mercury vapour sensor. Mercury is used extensively in laboratories, e.g. in manometers, electrodes and thermometers, and loss of mercury can therefore result in toxic levels of the material in these environments. Thus, a good sensor for mercury is highly desirable. A device containing a gold-plated quartz crystal has been use to sense mercury, as the latter readily forms an amalgam with gold. Heating to 150°C reverses this reaction.

Various acidic coatings can be used for the detection of ammonia, with ascorbic acid, L-glutamic acid and pyridoxine (vitamin B_6) hydrochloride having all been employed for this purpose. The latter gives a particularly reversible binding with ammonia, plus a high selectivity for this gas. However, other basic analytes, such as amines, can act as interferants.

SAQ 7.2

How is the piezo-electric effect used in sensors?

7.1.3 Biosensor Applications

Much more selectivity is obtained from biosensing methods. Thus, formaldehyde can be detected by coating the piezo-electric crystal with formaldehyde dehydrogenase/NAD^+ as the selective layer. The enzyme is present in the dry state:

$$CH_2O + H_2O + NAD^+ \xrightarrow{\text{FDH}} NADH + HCO_2H + H^+$$

The enzyme and the glutathione cofactor were immobilized on a 9 MHz quartz crystal with glutaraldehyde cross-linking.

Organophosphorus pesticides can be detected by the use of metal salts such as $FeCl_3$, $CuCl_2$ and $NiCl_2$. These will form complexes with such compounds and have been used to make piezo-electric sensors. In addition, cholinesterase enzymes will react selectively with such compounds as 'Malathion'; acetylcholine esterase is immobilized on to a quartz crystal with glutaraldehyde, with a 5 ppm concentration of water being needed to operate this sensor device.

Antibodies make ideal, i.e. highly selective and highly sensitive, coatings on piezo-electric crystals, e.g. 'anti-Parathion' can be used to determine 'Parathion', but again a constant humidity is required. Sometimes, immunosensors are too selective, and thus a more 'general' organophosphorus sensor is needed. Nevertheless, there are many applications in the medical field, where high selectivity is required, such as for immunoglobins, and for *Candida albicans*, a yeast-like fungus which can be found in humans.

7.1.4 The Quartz Crystal Microbalance

It is often more practical to use a differential-mode system, with two balanced crystals and oscillators, as shown in Figure 7.2. Such a system is known as a *quartz crystal microbalance* (QCM). This device is currently receiving

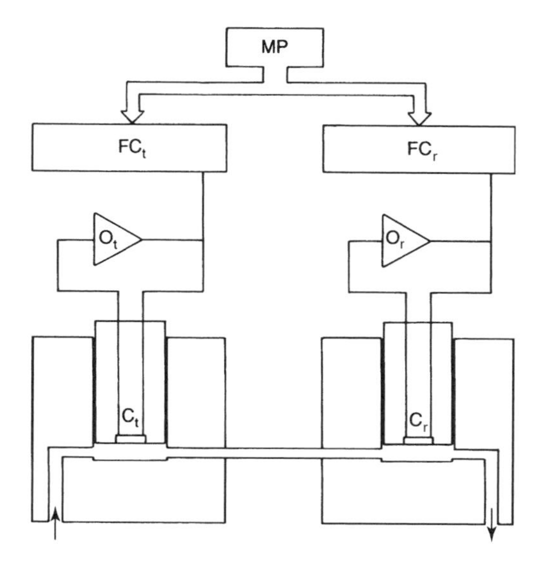

Figure 7.2 Schematic of a typical quartz crystal microbalance system, operating in the differential mode, with two balanced crystals and oscillators, used in electrogravimetric sensor analyses: C, crystal sensor; O, oscillating circuit; FC, frequency counter; MP, microprotector; the subscripts 'r' and 't' refer to reference and test, respectively. Reprinted from **Biosensors: Fundamentals and Applications** edited by A. P. F. Turner, I. Karube and G. S. Wilson (1987), by permission of Oxford University Press.

considerable attention. There have been problems with variability of balance between the two sections and hence a lack of sensitivity and poor signal-to-noise (S/N) ratios. A variation of this uses an amplification scheme – this is referred to as an amplified mass absorbent assay (AMISA). This latter system has been used to measure adenosine 5′-phosphosulfate reductase (APS). In this, an alkaline phosphatase antibody was bound to the microbalance surface, which was then exposed to 5-bromo-4-chloro-3-indolyl phosphate. This caused precipitation of an insoluble dephosphorylated dimer on the balance surface, thus enabling 5 ng cm^{-3} of APS to be detected.

Other applications have been to *Salmonella typhimurium* and to a DNA strand of a *Herpes simplex* virus.

A related device is the *electrochemical quartz crystal microbalance* (EQCM). A thin layer of a metal such as gold is plated on to the surface of a piezo-electric crystal, which is made the working electrode in an electrochemical cell. The EQCM will detect changes in the mass of material at the electrode, such as (a) adsorption/desorption of species at the monolayer level, (b) electrodeposition/electrodissolution at the electrode due to Faradaic redox processes, and (c) transfer of species between the solution and a surface-immobilizing film.

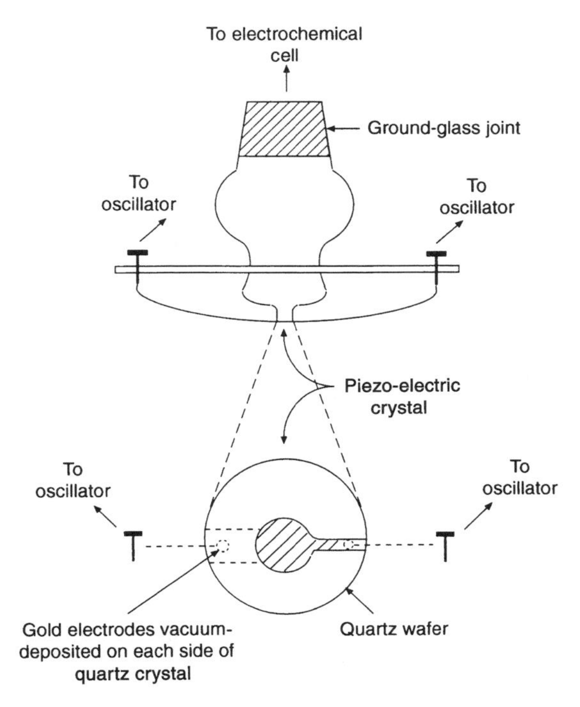

Figure 7.3 Schematic of the electrochemical quartz crystal microbalance. From Eggins, B. R., *Biosensors: An Introduction*, Copyright 1996. © John Wiley & Sons Limited. Reproduced with permission.

A typical EQCM is made from a 10 MHz AT-cut quartz crystal giving a sensitivity of 4 ng Hz^{-1}, as shown in Figure 7.3. It is preferable to compensate for changes in the properties of the solution, such as viscosity, which is affected by temperature, or density. Polymer-modified electrodes can be used, by monitoring the changes in mass measured by the EQCM. Some polymers used in this way include the redox polymers, polythionine and polyvinylferrocene.

SAQ 7.3

How is selectivity obtained in mass-sensitive sensors?

7.2 Surface Acoustic Waves

These are formed by using piezo-electric crystals, particularly lithium niobate ($LiNbO_3$), although such waves are not generated in the bulk of the solution,

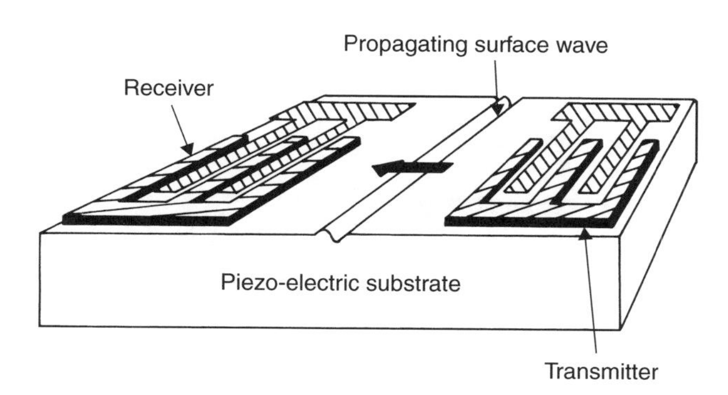

Figure 7.4 Schematic of a surface acoustic wave sensor device. From Zhang, D., Green, G. M., Flaherty, T. and Swallow, A., *Analyst*, **118**, 429–432 (1993). Reproduced with permission of The Royal Society of Chemistry.

but on the surface. A transmitter and receiver are positioned at each end of the crystal, as shown in Figure 7.4.

The transmitter and receiver usually consist of sets of interdigitated electrodes. A radiofrequency applied from the transmitter produces a mechanical stress in the crystal, so producing a Raleigh-type *surface acoustic wave* (SAW) which is received by the second set of electrodes and thus translated into an electrode voltage. The surface wave penetrates into the crystal to a depth of about one wavelength (rather like evanescent optical waves). Thus, a species immobilized on the surface will affect the transmission of the wave, unless the crystal is excessively thick. A number of variations on this technique exist.

7.2.1 Plate Wave Mode

This involves waves being reflected through the bulk of the piezoelectric crystal to an interdigitated transducer (IDT) on one of its surfaces.

7.2.2 Evanescent Wave Mode

This uses a substrate thickness less than the usual five acoustic wavelengths, i.e. down to about two to three acoustic wavelengths, in order to utilize the downward component of the (evanescent) wave. A ground plane is placed between the two IDT structures in order to prevent interference signals.

7.2.3 Lamb Mode

This uses a very small substrate thickness, compared to the wavelength, and the excitation frequency is much lower. Often, an analytically sensitive polymer layer is present on the lower surface, as shown in Figure 7.5.

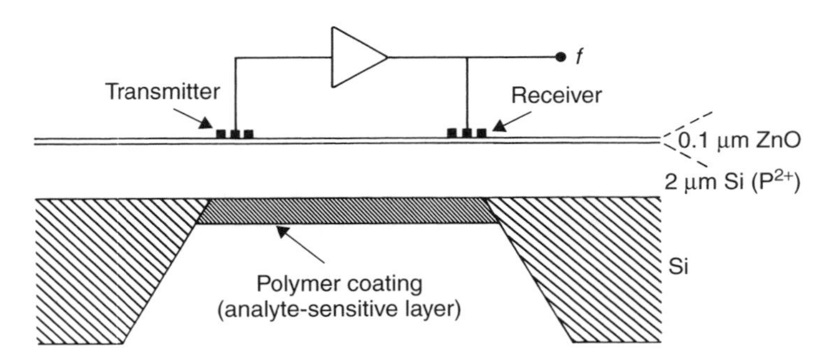

Figure 7.5 A transverse section through a Lamb mode surface acoustic wave sensor device. From Hall, E. A. H., *Biosensors*, Copyright, 1990. © John Wiley & Sons Limited. Reproduced with permission.

This mode offers a higher sensitivity and greater flexibility of sensor design. It has been used with a copper phthalocyanine coating for vapour analysis.

Although a considerable amount of work has been devoted to the study of the properties of these devices, often involving glycerol solutions, very few practical applications have been realized. Those systems studied include the influenza virus and DNA.

7.2.4 Thickness Shear Mode

This mode has been mainly applied to liquids. Extra variables are involved in this case, such as the viscosity, density and conductivity of the liquid being studied. Shear-mode devices generate only two types of analytical signals. First, thin films characterized by a shear modulus of elasticity give rise to standard Sauerbrey mass measurements. Secondly, capture at a liquid–solid shearing surface could lead to a differential signal associated with the introduction of a new material at the interface. Applications have been made of this device towards biosensor development in several ways.

Candida albicans yeasts were detected by means of the anti-*Candida* antibody, which was covalently bonded on to plated platinum electrodes. In addition, human IgG was measured on an AT-cut 9 MHz crystal, modified with protein A being immobilized on an oxidized palladium layer on the crystal surface. Shifts in frequency were ascribed to the affinity of protein A for human IgG.

DQ 7.1

Summarize the relative merits of the different modes of surface acoustic waves (SAWs) for use in sensors devices.

Answer

The ***basic SAW mode*** *uses Rayleigh surface waves generated by a radio-frequency wave applied to the surface of a piezo-electric crystal, which* *then interacts with the surface coating of an analyte species. The wave is detected by a second set of electrodes and converted into a voltage.*

In the **plate mode**, *waves are reflected through the bulk crystal to an interdigitated transducer.*

The **evanescent wave mode** *involves a lower substrate concentration, and can use the perpendicular component of the acoustic wave.*

In the **Lamb mode**, *there is a smaller substrate thickness and a lower excitation frequency. A polymer layer on the lower surface results in greater sensitivity and flexibility of design.*

The **thickness shear mode** *is used mainly for liquids, and partially depends on the viscosity, density and conductivity of the liquid being studied. Thin films characterized by the shear modulus of elasticity give Sauerbrey mass measurements. A differential mode variation can also be used.*

SAQ 7.4

What detection limits are possible with mass-sensitive sensors?

7.3 Thermal Sensors

7.3.1 Thermistors

A thermistor is a very sensitive device for measuring changes in temperature. Its operation is based on the change in electrical resistance with temperature of certain sintered metal oxides. These include BaO, CaO, or transition-metal oxides such as those of Co, Ni and Mn. There is usually a decrease in resistance of 4–7% per degree rise in temperature, with an accuracy of ±0.005°C. Such devices are usually constructed in the form of small glass beads and so can be regarded as miniaturized systems. Thermistors can be used to measure the small amounts of heat evolved in a chemical or biochemical reaction by employing a microcalorimeter. Selectivity can be achieved by carrying out the particular reaction close to the thermistor component, which thus only involves the analyte of interest. More usefully, this technique can be used to measure the enthalpy change in an enzymatic reaction, which, as we have seen earlier, imparts the selectivity. The method can be carried out with coloured or turbid solutions, when colorimetric methods are inapplicable. The enzyme, mixed with albumin, can be coated (immobilized) on to the thermistor by cross-linking with glutaraldehyde, (see Figure 7.6), so making it a discreet sensor. A second thermistor coated only with albumin, acts as a reference. The response is a resistance change, thus giving an electrical signal.

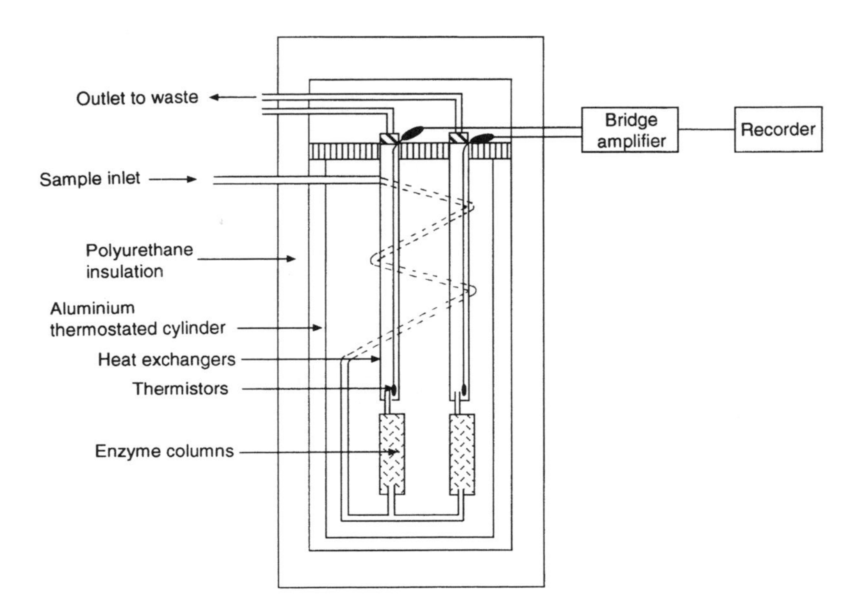

Figure 7.6 Schematic of a typical thermistor device, as used as a biosensor in enzymatic reactions. Reprinted from **Biosensors: Fundamentals and Applications** edited by A. P. F. Turner, I. Karube and G. S. Wilson (1987), by permission of Oxford University Press.

The signal obtained is distinctly non-linear, as represented by the following equation:

$$R_T = A\exp\,(\beta/T), \text{ or } R_{T1} = R_{\text{T0}}\exp\,[\beta(1/T_1 - 1/T_0)]$$

where β usually has a value of about 3000. Typically, R is about 10 kΩ at 300 K, with the slope of the corresponding plot being 1 kΩ/°C. The usual temperature range of applicability is -100 to 200°C.

However, the response can be made sensibly linear over a small temperature range. In any case, this non-linearity can be compensated for by using a microprocessor measuring device, although this approach has not received any wide application to date.

Two examples of the application of microcalorimetry to enzyme reactions are shown below. The amount of heat evolved depends both on the amount of analyte and on the enthalpy of the reaction:

$$\text{glucose} + O_2 + H_2O \xrightarrow[\Delta H\,=\,-80\text{ kJ mol}^{-1}]{\text{GOD}} \text{gluconic acid} + H_2O_2$$

$$CO(NH_2)_2 + H_2O \xrightarrow[\Delta H\,=\,-6.6\text{ kJ mol}^{-1}]{\text{urease}} CO_2 + 2NH_3$$

It can be seen that the glucose sensor is far more sensitive than the urea sensor, because of the much larger enthalpy. Glucose can be determined down to concentrations of 2 mM by using such a sensor.

> **SAQ 7.5**
>
> What is a thermistor?

7.3.2 Catalytic Gas Sensors

Catalytic sensors are used particularly for detecting flammable gases. Their operation involves the controlled combustion of a flammable gas in air and the measurement of the quantity of heat liberated in the process, for example:

$$CH_4 + 2O_2 \xrightarrow[\Delta H = -800 \text{ kJ mol}^{-1}]{} CO_2 + 2H_2O$$

To speed up the process, a catalyst, such as platinum, is used. A coil of platinum wire is used as a heater to maintain the gas mixture at the combustion temperature, while the resistance of the platinum wire also acts by responding to changes in temperature. An electric current is passed through the platinum coil to heat it, and hence the gas, to combustion temperature. The heat evolved increases the temperature of the wire, and thus its electrical resistance, which is then measured. With methane, the heat evolved is quite large, so this is a very sensitive method of analysis for this hydrocarbon. Such a sensor device is termed as operating in the *non-isothermal* mode.

An alternative approach is to use the electronic circuitry in a feed-back device to decrease the current, so as to maintain the temperature of the wire constant, by compensating for the heat evolved in the reaction. This is known as the *isothermal mode* of operation. The change in the current is measured and is related to the heat evolved by the combustion of the gas, and hence to the quantity of gas present.

A temperature of ca. 1000°C is required when using platinum as a catalyst. Other catalysts, such as palladium or rhodium, operate at lower temperatures and are therefore preferable.

An alternative form of catalytic gas sensor is the *pellister*. The principle is the same as above, except that here the platinum wire resistance thermometer/heating wire is embedded in a ceramic bead. The catalyst layer, usually palladium, is coated on the surface of the bead (as shown in Figure 7.7). Such a system will operate at a temperature of 500° for various gases, e.g. methane.

A particular problem with catalytic sensors is that the catalyst can be poisoned by organosulfur or organophosphorus compounds and alkyllead derivatives. In order to resolve this, the bead of the sensor can be made of porous alumina in which the catalyst is mixed (as shown in Figure 7.8).

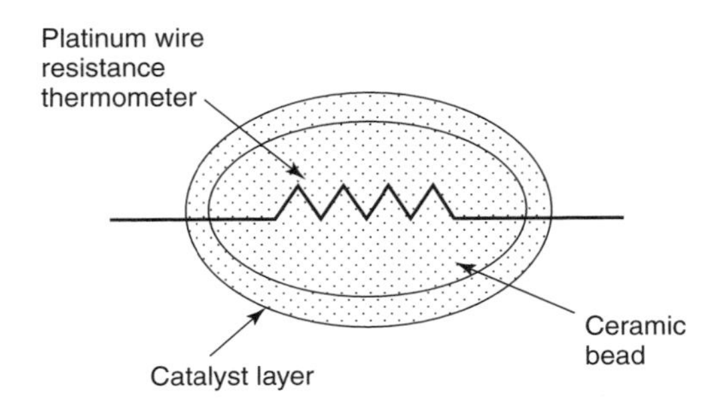

Figure 7.7 The pellister catalytic gas sensor. © R. W. Catterall 1997. Reprinted from **Chemical Sensors** by R. W. Catterall (1997), by permission of Oxford University Press.

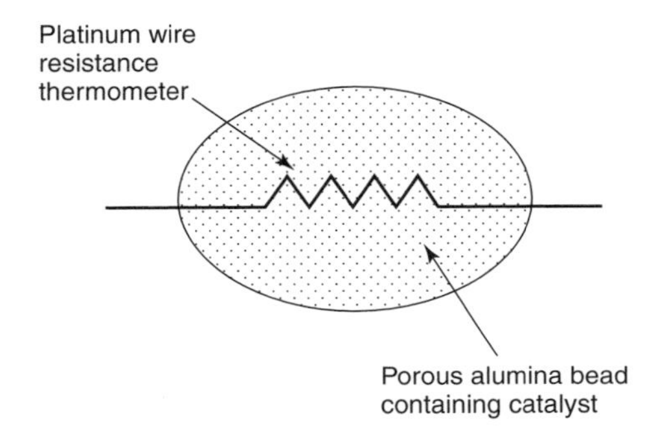

Figure 7.8 A 'poison-resistant' pellister gas sensor. © R. W. Catterall 1997. Reprinted from **Chemical Sensors** by R. W. Catterall (1997), by permission of Oxford University Press.

These gas sensors are usually operated in pairs, one of which contains no catalyst. They then form separate arms in a Wheatstone bridge circuit, employing an arrangement such as that shown in Figure 7.9.

DQ 7.2

Discuss the application of thermal sensors to gas analysis.

Answer

Thermal sensors are applied mainly to the analysis of combustible gases, particularly those found in mixtures with air, such as methane and carbon monoxide, and also to volatile flammable liquids. Although the technique is fairly non-selective, it can be used in places where only single known gases are expected, such as methane in mines, or volatile liquids in chemical factories. The principle for all of these sensors is to burn the gas in air and then measure the amount of heat emitted. This is really the same

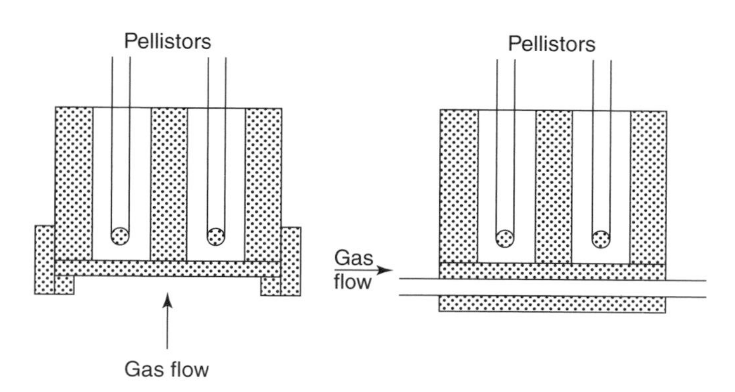

Figure 7.9 Schematic of a bridge arrangement of two pellister gas-sensing devices. © R. W. Catterall 1997. Reprinted from **Chemical Sensors** by R. W. Catterall (1997), by permission of Oxford University Press.

as discussed above for thermistors, except that in this case the reaction is combustion, rather than enzymatic. All use some form of catalyst to enhance the rate of combustion. The most sensitive device is the ***pellister****.*

These thermal sensors, many of which operate on the piezo-electric principle, often have coatings on the sensor element which will selectively react with the gas being detected. Some examples are as follows:

- *Water vapour – a film of hygroscopic material, such as silica gel or molecular sieves*
- *Sulfur dioxide – organic amines*
- *Ammonia – several materials, including pyridoxine hydrochloride*
- *Hydrocarbons – 'Carbowax 550', although not very selective*
- *Hydrogen sulfide – silver, copper or lead acetates*
- *Mercury vapour – gold plating, which amalgamates the mercury*
- *Carbon monoxide – mercury(II) oxide, which is reduced to mercury vapour; the latter is then detected as above*

7.3.3 Thermal Conductivity Devices

These measure the thermal conductivity of a gas and are typically used in gas chromatography (GC or GLC). A filament, made of tungsten, tungsten/rhenium alloy, or nickel/iron alloy, is heated electrically to about 250°C. Heat is lost to the surroundings at a rate which is dependent on the thermal conductivity of the surrounding gas. The change in temperature of the filament is determined by the change in electrical resistance in the same way as for other calorimetric sensors. However, such devices are not used to measure chemical reactions and therefore can be safely employed in an inert environment for monitoring flammable gas residues. They can also be used for measuring inert gases themselves, such as

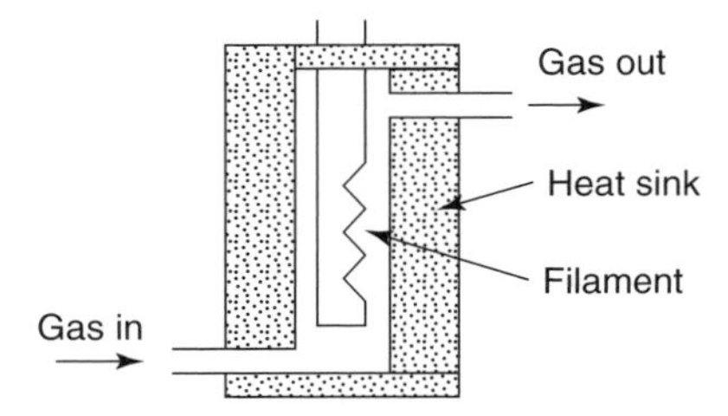

Figure 7.10 Schematic of a thermal conductivity detector for use in gas chromatographic analysis. © R. W. Catterall 1997. Reprinted from **Chemical Sensors** by R. W. Catterall (1997), by permission of Oxford University Press.

nitrogen, argon, helium or carbon dioxide. A schematic diagram of a typical thermal conductivity device used in GC analysis is shown in Figure 7.10.

SAQ 7.6

What is a pellister?

SAQ 7.7

Which three thermal processes are used in sensor devices?

Summary

This chapter describes the principles of piezo-electricity, piezo-electric sensors, and their application to sensing gases and to biosensors. Radiofrequency (RF) oscillations and inductance–capacitance circuits are described, along with the use of piezo-electric crystals to control the RF. Anisotropic crystals and the piezo-electric effect, plus the effect of mass changes on the crystal frequency – the Sawerbury equation – are outlined. Further developments of the piezo-electric effect to the quartz crystal microbalance and the electrochemical quartz crystal microbalance are described. Modification of the piezo-electric effect to produce surface acoustic waves is given, along with its variations, i.e. the plate wave mode, the evanescent mode, the Lamb mode and the shear mode. Applications of surface acoustic waves in vapour analysis and in biosensors are described.

Three main thermal sensing devices are also discussed, i.e. calorimetric, catalytic and thermal conductivity. The construction and operation of a thermistor is presented, along with its application in biosensors. In addition, the principles and applications of the catalytic gas sensor and the pellister are described, as well as the use of thermal conductivity devices as gas chromatographic detectors.

Further Reading

Bruckenstein, S. and Shay, S., 'Applications of the quartz crystal microbalance in electrochemistry', *J Electroanal. Chem.*, **188**, 131–135 (1985).

Christiansen, P. A. and Hamnett, A., 'The electrochemical quartz microbalance', in *Techniques and Mechanisms in Electrochemistry*, Blackie, Glasgow, UK, 1994, pp. 204–208.

Clark, D. J., Blake-Coleman, B. C. and Calder, M. R., 'Principles and potential of piezo-electric transducers and acoustical techniques', in *Biosensors: Fundamentals and Applications*, Turner, A. P. F., Karube, I. and Wilson, G. S. (Eds), Oxford University Press, Oxford, UK, 1987, pp. 551–572.

Danielson, B. and Mosbach, K., 'Theory and application of calorimetric sensors', in *Biosensors: Fundamentals and Applications*, Turner, A. P. F., Karube, I. and Wilson, G. S. (Eds), Oxford University Press, Oxford, UK, 1987, pp. 575–597.

Thompson, M., Kipling, A. L., Duncan-Hewitt, W. C., Rajaković, L. V. and Čavić-Vlasak, B. A., 'Thickness-shear-mode acoustic wave sensors in the liquid phase: A review', *Analyst*, **116**, 881–890 (1991).

Walton, P. W., Gibney, P. M., Roe, M. P., Lang, M. J. and Andrews, W. J., 'Potential biosensor systems employing acoustic impulses in thin polymer films', *Analyst*, **118**, 425–428 (1993).

Zang, D., Green, G. M., Flahery, T. and Shallow, A., 'Development of inter-digitated acoustic wave transducers for biosensor application', *Analyst*, **118**, 429–432 (1993).

Chapter 8

Specific Applications

Learning Objectives

- To see the application of a number of the sensor techniques presented in this text to five different analyses.
- To appreciate the structure and performance of the Medisense 'ExacTech' biosensor for the determination of glucose in blood.
- To see how complexation and differential pulse polarography are combined to measure low concentrations of copper(I) in water.
- To understand the developments towards a 'laboratory on a chip' – using arrays of sensors to determine mixtures of analytes.
- To see specifically the use of ISFET sensors for the simultaneous determination of four different metal ions in blood.
- To understand how the combination of an immunoassay linked via enzyme processes to a bioluminescent assay can determine pmol amounts of explosive materials.
- To appreciate the use of a broad-spectrum enzyme available in plant tissues in the amperometric analysis of flavanols in beers.

DQ 8.1

What are the main factors to consider when devising a new sensor?

Answer

(a) Any special criteria for the application.
(b) Decide on the selective element.
(c) Select the transducer.
(d) Decide on the method of immobilization.
(e) The performance factors required.

(*f*) *Construction of the sensor.*
(*g*) *Operation of the sensor*
(*h*) *Testing of the sensor.*

8.1 Determination of Glucose in Blood – Amperometric Biosensor

8.1.1 Survey of Biosensor Methods for the Determination of Glucose

It is said that about half of the research papers published on biosensors are concerned with glucose. In addition to its metabolic and medical importance, this material provides a good standard compound on which to try out possible new biosensor techniques. There are, in fact, a number of different ways of determining glucose just by using electrochemical transducers. Reference to this has already been made under other transducer modes in Chapters 5 and 6 earlier.

Figure 5.18 shown above illustrates the overall pattern for assays which use these transducers. It can be seen from this figure that there are several different ways in which glucose may be determined, although glucose oxidase (GOD) lies at the centre of them all.

There are of course many other systems that can be used to determine glucose, employing different enzymes, such as glucose dehydrogenase (GDH), or different transducers, including thermal and photometric devices.

8.1.2 Aim – to provide a simple, portable, sensor for use at home by diabetics for regular monitoring of the level of glucose in their own blood

8.1.2.1 Special Criteria for this Application

For use by patients at home, it must be simple, reliable and cheap.

8.1.2.2 Decide on the Selective Element

Glucose oxidase is an inexpensive, readily available enzyme obtained from *Aspergillis niger*. It is stable over a long period of time, especially when sealed in foil.

8.1.2.3 Select the Transducer

An electrochemical transducer, particularly an amperometric type, is cheap, reliable and will give a direct read-out to a liquid crystal display (LCD). Most commercial glucose sensors use glucose oxidase coupled to a mediator (usually a ferrocene derivative) as the electron donor, rather than oxygen.

8.1.2.4 Decide on the Method of Immobilization

For 'long life', a covalent-bonding method is best. The original biosensors on which the Medisense 'ExacTech' sensor was based used a graphite foil coated with dimethylferrocene as the mediator. The oxidase was immobilized by reaction with 1-cyclohexyl-3-(2-morpholinoethyl)carbodiimide *p*-methyltoluenesulfonate. The Medisense 'ExacTech' device appears to use a screen-printed technique.

8.1.2.5 The Performance Factors Required

- Blood glucose range of 1.1–33.3 mM
- Precision of ±3–8%
- Test time of −30 s
- Lifetime of sensor – test-electrode strip to last at least six months in sealed foil

8.1.2.6 Construction of the Sensor

The device shown in Figure 8.1 has dimensions of 53 × 90 mm and weighs 40 g. It uses a non-replaceable 3 V battery that lasts for about 4000 tests. It is marketed at about £35[†] and is replaced free when the battery runs out.

The electrode strip is shown in more detail in Figure 8.2. This is used just once to ensure optimum reproducibility. These strips cost about 40 p each.[†] Miniaturization is achieved by immobilizing the enzyme and mediator on to a screen-printed strip, such as the one shown in Figure 8.1. One electrode contains the immobilized enzyme–mediator, while the other is a silver–silver chloride reference electrode. (Silver–silver chloride 'ink' can be screen-printed on to the appropriate electrode.) The electrode strip has two metal contacts at one end which are connected to the measuring device.

8.1.2.7 Operation of the Sensor

A drop of blood is placed on the strip across the two electrodes, making contact with both and thus obviating the need for a cell container. No supporting electrolyte is needed, and no de-gassing is involved.

The switch on the measuring device is then pressed. The timing is automatic and is usually displayed on the LCD read-out. After 30 s, the current is read and the reading is then converted into a measure of glucose in the blood and displayed.

The transducer is an electrode that is set to a potential (versus a reference electrode) at which the reduced ferrocene derivative may be re-oxidized, i.e. it is used in the amperometric mode. The oxidation current is measured for a fixed time (usually 30 s) and is then recorded and displayed on the LCD.

[†] Price as at September 2001.

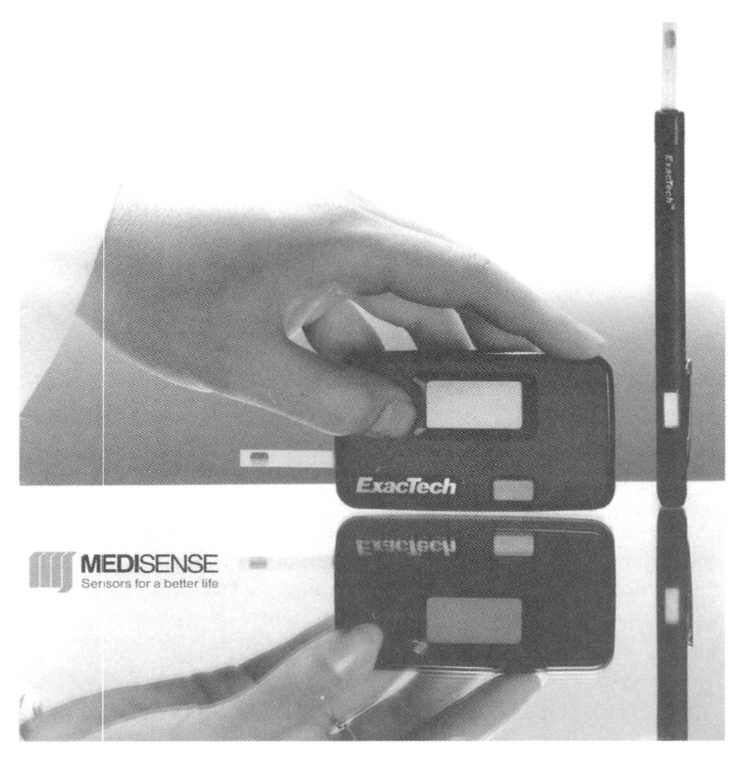

Figure 8.1 The Medisense 'ExacTech' biosensor device. Reproduced by permission of Medisense, Birmingham, UK.

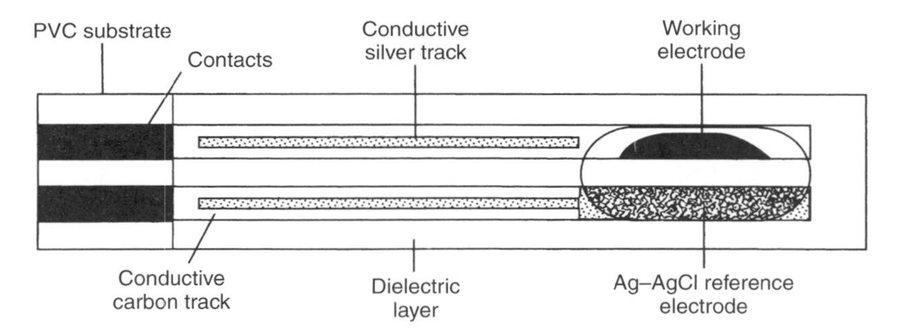

Figure 8.2 Schematic of the 'ExacTech' biosensor disposable electrode strip. From Hilditch, P. I. and Green, M. J., *Analyst*, **116**, 1217–1220 (1991). Reproduced with permission of The Royal Society of Chemistry.

8.1.2.8 Testing of the Sensor

The following shows test data provided by Medisense.

Precision Three blood samples with three different glucose levels were tested, with 20 readings being taken using electrode test strips from the same batch. The results obtained are shown in the table below.

Parameter	Sample		
	1	2	3
Number of readings	20	20	20
Mean concentration (mM)	3.0	4.8	15.5
Standard deviation	0.24	0.19	0.52
Coefficient of variation (%)	8.1	3.9	3.3

Accuracy A clinical study was carried out, comparing the 'ExacTech' sensor with the blood glucose-monitoring device for laboratory use from Yellow Springs Instruments (YSI Model, No. 23AM). The correlation over 200 readings was given by the following linear regression:

$$y = 1.041x + 0.271 \text{ mM}$$

with a correlation coefficient of 0.985.

Patient use A study compared the results obtained from a group of nurses and medical technicians with those obtained from a number of patients. The similarity in the findings showed that the 'ExacTech' device performed equally well in the hands of patients as it did with trained professionals.

SAQ 8.1

How could a glucose analyser be used for continuous monitoring of a patient's blood glucose levels?

8.2 Determination of Nanogram Levels of Copper(I) in Water Using Anodic Stripping Voltammetry, Employing an Electrode Modified with a Complexing Agent

8.2.1 Background to Stripping Voltammetry – Anodic and Cathodic

During cyclic voltammetry, the concentration of the reduced species builds up at the electrode surface. This process can be improved if the material is adsorbed on to the surface in some way. The simplest method when using a (hanging) mercury-drop electrode is when the reduced species is a metal and forms an amalgam with the mercury. This effectively acts as a concentration process, which can be enhanced by holding the reduction potential at a value just past the peak potential, as shown in Figure 8.3(b).

After a certain period of time (say 1 min), the potential scan is reversed and so the material on the electrode surface is re-oxidized, thus giving a strongly enhanced inverted oxidation peak current, as shown in Figure 8.3(a).

With the development of solid electrodes, analytes were found which could be oxidized during the forward (now positive) scan and adsorbed on to the surface of the electrode in greater concentrations. These could then be re-reduced (cathodically stripped), so giving a greatly enhanced inverted reduction peak current.

While the oxidation (or reduction) potential provides some selectivity, this can be improved, as can the sensitivity, by coating the electrode surface with a material which selectively complexes with the specific analyte. This is the basis of the technique for analysing low concentrations of metal ions such as copper(I). A very simple way to achieve this is to mix the complexing agent with the paste of a carbon paste electrode (or with the carbon ink for a screen-printed electrode).

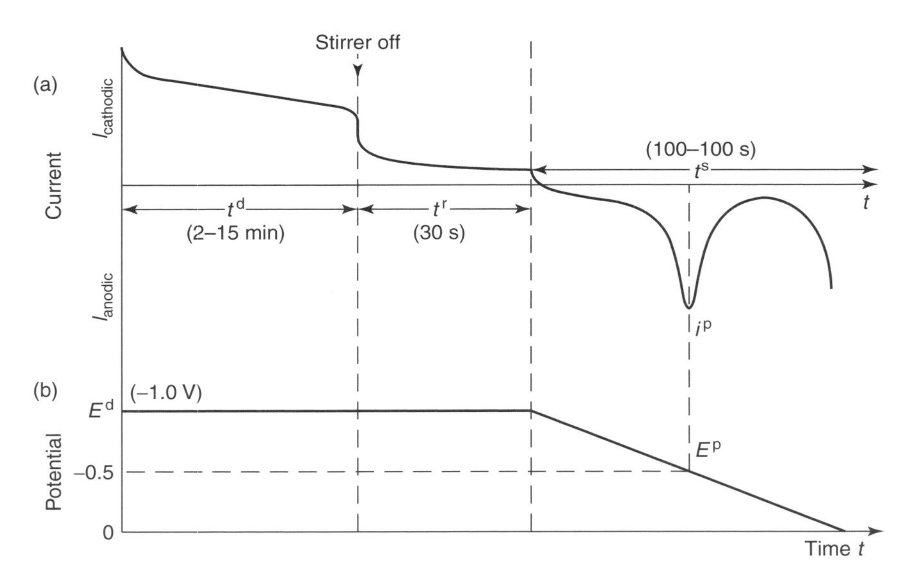

Figure 8.3 (a) Typical anodic stripping voltammogram obtained when using a hanging-mercury-drop electrode. (b) Corresponding variation with time of the potential applied to the electrode during the stripping voltammetry procedure. From Bard, A. J. and Faulkner, L. R., *Electrochemical Methods and Applications*, 2nd Edition, © Wiley, 2000. Reproduced by permission of John Wiley & Sons, Inc.

8.2.2 Aim – to make an ultra-sensitive sensor that will detect ppb levels of copper(I) ions

8.2.2.1 Special Criteria for this Application

Measurement of very low concentrations of ions in water samples with a high degree of selectivity.

8.2.2.2 Decide on the Selective Element

Ion-selective electrodes are not sensitive enough for this application. However, 2,9-dimethyl-1,10-phenanthroline (2,9-DMP) will form a complex with copper(I). When this is incorporated into a carbon paste electrode, there is no response from Cu(II), Fe(III), Ni(II), Co(II), Zn(II) or Pb(II). Thus, selectivity is achieved via a combination of the selectivity of the complexing agent and the selectivity of the voltammetric potential.

8.2.2.3 Select the Transducer

A voltammetric electrode is used in this case.

8.2.2.4 Decide on the Method of Immobilization

The carbon paste entrapment method is very suitable for this particular application.

8.2.2.5 The Performance Factors Required

- A linear range of 0.3–10 μM (±5%), with a detection limit of 5 nM
- A concentration time of 1 s to 3 min

8.2.2.6 Construction of the Sensor

Graphite powder was mixed with the 2,9-DMP, Nujol was added, and the whole mixed into a paste. The electrode holder was made from a plastic syringe tube, with an electrode area of 0.02 cm^2. Cyclic voltammetry was carried out with a BAS CV-1B potentiostat, while for differential pulse experiments an IBM Model EC225 differential pulse polarograph was used.

8.2.2.7 Operation of the Sensor

The electrode was pre-treated with 1×10^{-3} M Cu(I) solution, washed and then kept at a potential of 0.35 V for 10 min. The sample was contained in a 40 ml cell with a pH 6.0 acetate buffer. A saturated-calomel reference electrode and a platinum wire counter-electrode were used. Deposition of the copper(I) was carried out chemically by immersion of the chemically modified electrode for various times in the sample solution. There is no pre-electrolysis step required in this method. Then, differential pulse polarography was carried out, using a 20 s pulse and a sweep rate of 10 mV s^{-1}.

8.2.2.8 Testing of the Sensor

This method was tested against a National Bureau of Standards (NBS) standard containing 21.9 ± 0.4 μM of copper(I), using both calibration and standard addition methods, according to the data shown in the table below.

Method	Number of determinations	Deposition time (s)	Copper found (ppb)		SD[a]
			Mean	Range	
Calibration	6	8	24.7	14.8–34.8	6.5
Standard addition	4	8	21.8	19.3–24.0	2.0

[a]SD, standard deviation.

Deposition times in the range from 1 s to 3 min were used. With the latter, a detection limit of 5.0 nM was obtained, with a signal-to-noise ratio (S/N) of 2:1. Silver showed some interference, but at a level of 9.8 ppb Ag(I) there was no problem.

SAQ 8.2

To what extent is this a cathodic stripping voltammetric analysis?

8.3 Determination of Several Ions Simultaneously – 'The Laboratory on a Chip' *(Application to the analysis of cations in blood using an ISFET device)*

8.3.1 Chemiresistors

Sensors based purely on variation of the resistance of the sensor in the presence of an analyte generally lack selectivity. However the use of an array of several sensing elements, each with a slightly different resistance response, can be used to detect quite complex mixtures. Each sensing element can be coated with a different conducting polymer, or made from sintered metal oxides. These arrays can develop a unique signature for each analyte. Such signatures can be obtained from mixtures of analytes and can be used to test flavourings in beers and lagers (and perhaps wines), and to test the aromas of coffee blends. This system is often referred to as an 'electronic nose'. Analysis of the signal output from an array of maybe 12 to 20 sensor elements uses neural network analysis which simulates brain function activity. The data obtained can be used to construct non-parametric, non-linear models of the array response. A number of these devices have been commercialized.

8.3.2 Sensor Arrays and 'Smart' Sensors

Integrated signal processing eliminates electronic noise, stray capacitance and electromagnetic interference, thus improving the signal-to-noise ratio (S/N). Some systems can carry out self-testing and self-calibration, i.e. detect and diagnose problems (as a PC does – usually!). There can also be automatic calibration and temperature correction.

8.3.2.1 Sensor Arrays

Several sensors are connected to an 'intelligent' processor which takes the individual outputs and analyses the response patterns, extracting analytically useful information. For example, we could have 10 identical pH electrodes in an array. The response of those probes that do not match the majority response can provide self-diagnosis, or each could monitor acidity in different locations within the sample.

Arrays can be made which respond to different chemical species. Partially selective sensors each have some discriminating power between analyte and interferent, although the outputs from each are different. Different modes of transduction can be used in this way, including QCMs, MIS devices, FETs and amperometric sensors. A schematic representation of a chemically intelligent sensor array is shown in Figure 8.4.

An application using ISEs for the ions Ca, Na and K involves an efficient four-electrode array of three highly selective electrodes and one sparingly selective

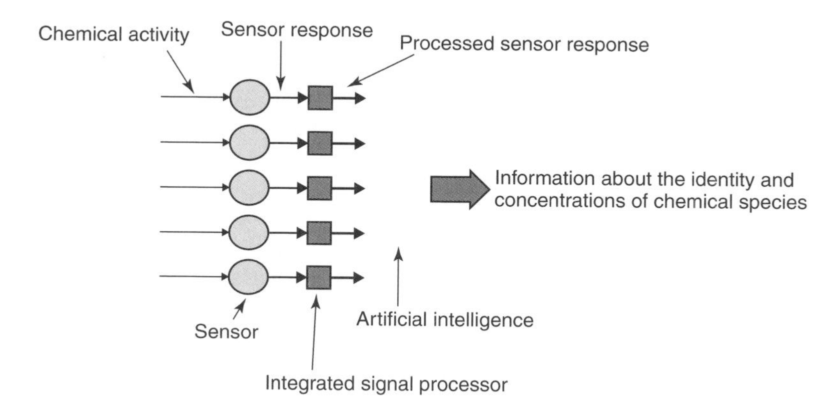

Figure 8.4 Schematic representation of a chemically intelligent sensor array. From Diamond, D. (Ed.), *Principles of Chemical and Biological Sensors*, © Wiley, 1998. Reproduced by permission of John Wiley & Sons, Inc.

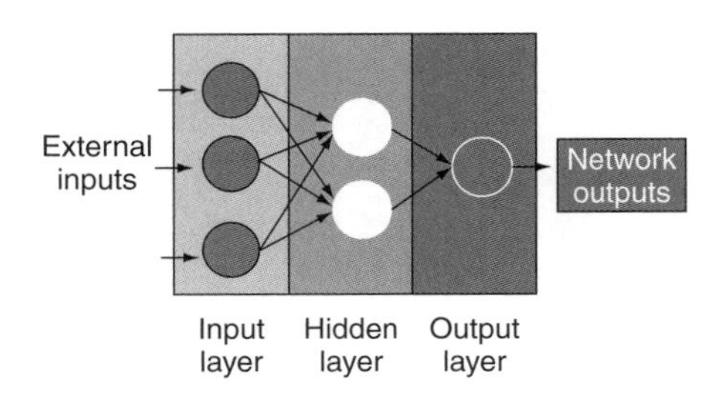

Figure 8.5 Schematic representation of a neural network. From Diamond, D. (Ed.), *Principles of Chemical and Biological Sensors*, © Wiley, 1998. Reproduced by permission of John Wiley & Sons, Inc.

electrode. The array response was operated as a detector in flow-injection analysis (FIA), and was modelled on the Nicholskii–Eisenmann equation using experimental data from calibration experiments selected according to a factorial experimental design. Non-linear optimization methods were used to evaluate the standard cell potential, slope, and selectivity coefficients. This deterministic approach produced better results than those obtained via the 'fuzzy logic' evaluation of data from neural network analysis.

This array, a so-called 'electronic nose', operates via a neural network configuration (see Figure 8.5).

A range of sensing elements could be used for such an application. Polymer-coated sensor wires with different resistor responses have already been described under chemresisitors in Section 8.3.1 above.

8.3.3 Background to Ion-Selective Field-Effect Transistors

The operation of FETs has been described above in Chapters 2 and 5. The ionophore is the most useful type of ion-selective polymer to use in FET devices (and also with ISEs). However, with FETs there are also some special responses for H^+. The first pH ISFET used the bare insulator gate as the ion-sensitive layer, but SiO_2 was not very effective, owing to its easy hydroxylation. However, silicon nitride (Si_3N_4)-gate devices are not hydrolyzed and are highly selective to H^+ ions, with a response of 50–60 mV per decade (pH). TiO_2 and Ge show a similar response. These semiconductor materials can be handled by the same techniques as used for FET chip fabrication. For other ions, such techniques are less successful. However, for Na^+ ions, borosilicate glass can be deposited in the gate region by integrated circuit processes.

Polymer membranes have been used successfully for K^+ determinations, by incorporating valinomycin and crown ethers and with *p*-(1,1,3,3-tetramethylbutyl)phenyl phosphoric acid for Ca^{2+}. The responses are <40 mV per decade, unless the membrane is thicker than 100 μm.

Heterogeneous membranes have had some success, with the best being polyfluorinated phosphazine (PNF) mixed with silver salts, in particular silver chloride

(75%) with PNF (25%). The response for Cl^- was 52 mV per decade. Changing the mixture and including Ag_2S or AgI can adjust the selectivity to favour a particular ion.

8.3.4 Aim – to develop a miniaturized sensor that will simultaneously determine several ions in a human blood sample

8.3.4.1 Special Criteria for this Application

Medical applications for determining multiple analytes, particularly ions.

8.3.4.2 Decide on the Selective Element

Ion-selective elements should be used, i.e. a Si_3N_4 bare gate for H^+, a glass coating for Na^+, a phosphoric acid derivative ionophore in a polymer membrane, such as PVC, for Ca^{2+}, and valinomycin, in a PVC membrane, for K^+.

8.3.4.3 Select the Transducer

This should be a field-effect transistor with four gates giving an output potential which is proportional to the logarithm of the concentration of each ion.

8.3.4.4 Decide on the Method of Immobilization

Various thin-layer techniques, compatible with integrated circuit processes, can be employed.

8.3.4.5 The Performance Factors Required

In human blood, the required ranges are 7.35–7.45 for pH, with concentrations of 2.2–2.55 mM (Ca^{2+}), 3.5–5.0 mM (K^+) and 135–142 mM (Na^+). The main problem here is to avoid interference between the relatively high levels of sodium ions and the lower levels of calcium, potassium and hydrogen ions.

8.3.4.6 Construction of the Sensor

This involved an array of five ISFETs, with one being selective for each ion and one acting as a reference, all mounted on the same semiconductor chip (see Figure 5.26 earlier). Each ISFET has an own associated integrated circuit, similar to that shown above in Figure 2.27, giving a voltage output to an LCD display.

8.3.4.7 Testing of the Sensor

This sensor worked satisfactorily under laboratory conditions but with whole blood (medical conditions) there were problems with the sodium analysis due to the sensitivity and selectivity of the Na^+ membrane. The range and environment of the Na^+ in whole blood thus render such a device unsuitable in this situation.

SAQ 8.3

How could a photometric sensor be devised for the simultaneous measurement of these ions?

8.4 Determination of Attomole Levels of a Trinitrotoluene–Antibody Complex with a Luminescent Transducer

8.4.1 Background to Immuno–Luminescent Assays

8.4.1.1 Bioluminescence

Certain biological species, principally the firefly, can emit luminescence. This originates in a group of substances of varied structure known as the *luciferins* (see Figure 2.39 above).

The enzyme-catalysed oxidation of luciferin results in luminescence, as follows:

$$\text{luciferin} \xrightarrow{\text{luciferase, } O_2} \text{oxyluciferin} + h\nu \text{ (562 nm)}$$

Some of the luciferase reactions 'couple' with cofactors such as ATP, FMN and FADH, for example:

$$\text{ATP} + \text{luciferin} + O_2 \xrightarrow{\text{luciferase}} \text{AMP} + \text{PP} + \text{oxyluciferin} + CO_2. + H_2O + h\nu$$

This reaction is very sensitive down to femtomole (fmol) concentrations.

Bacterial luciferases do not involve luciferins but form an excited complex with reduced flavins such as FMNH as follows:

$$FMNH_2 + O_2 + RCHO \longrightarrow FMN + RCOOH + H_2O + h\nu \text{ (478–505 nm)}$$

Most analytical reactions involve NADH, as shown earlier in Figure 2.40.

8.4.2 Aim – to construct a sensor that will monitor sub-ppb levels of explosive residues in environmental situations

8.4.2.1 Special Criteria for this Application

To detect very low levels of explosive residues in environmental situations, for example, those leached from old ammunition or mine-working dumps.

8.4.2.2 Decide on the Selective Element

An antibody, which will be specific for trinitrotoluene (TNT) and so will not detect other explosives. Used to find residues in water from mine workings, and traces on hands of persons who have handled explosives.

8.4.2.3 Select the Transducer

A bioluminescent sensor will be suitable as this gives the most sensitive response (see Chapter 2 above).

8.4.2.4 Decide on the Method of Immobilization

The antibody for TNT is immobilized on sepharose.

8.4.2.5 The Performance Factors Required

A sensitivity down to 10^{-15} M is needed.

8.4.2.6 The Read-Out Mode

A light detector converts the response into a current via a photomultiplier.

8.4.2.7 Operation of The Sensor

This assay is a multi-step process, involving not only the antibody but also enzymes, which introduce an amplification effect:

(a) The TNT–antibody species is immobilized on sepharose and incubated with the sample of TNT.
(b) This is then incubated with TNT which has been labelled with the enzyme G-6-PDH (glucose-6-phosphate dehydrogenase), which competes with the sample for the immobilized TNT–antibody.
(c) The sepharose is then washed to remove unbound TNT and an assay is carried out for the TNT–enzyme label.
(d) From the amount of labelled TNT, the amount of original TNT sample can be determined. The label is the enzyme which catalyses the reduction of NAD^+ to NADH, which is then used to reduce FMN to $FMNH_2$, employing oxido-reductase.
(e) Finally, the $FMNH_2$ reacts with oxygen and bacterial luciferase to emit light, which is determined photometrically.

The reaction process for this assay is shown in Figure 8.6.

Thus, this process uses an antibody as the selective material, there is enzyme amplification, and finally the response is by photometric bioluminescence, so resulting in ultra-sensitivity.

A major problem with this technique is that enormous care has to be taken to avoid contamination of any of the materials or equipment with traces of sample. Failure to do this led to, for example, the 'Birmingham Six', suspects after the IRA bombing in Birmingham, being jailed for a long time, until eventually it was decided that the *positive* test result could have been due to contamination at some stage in the analysis by the forensic scientists who carried out the assay.

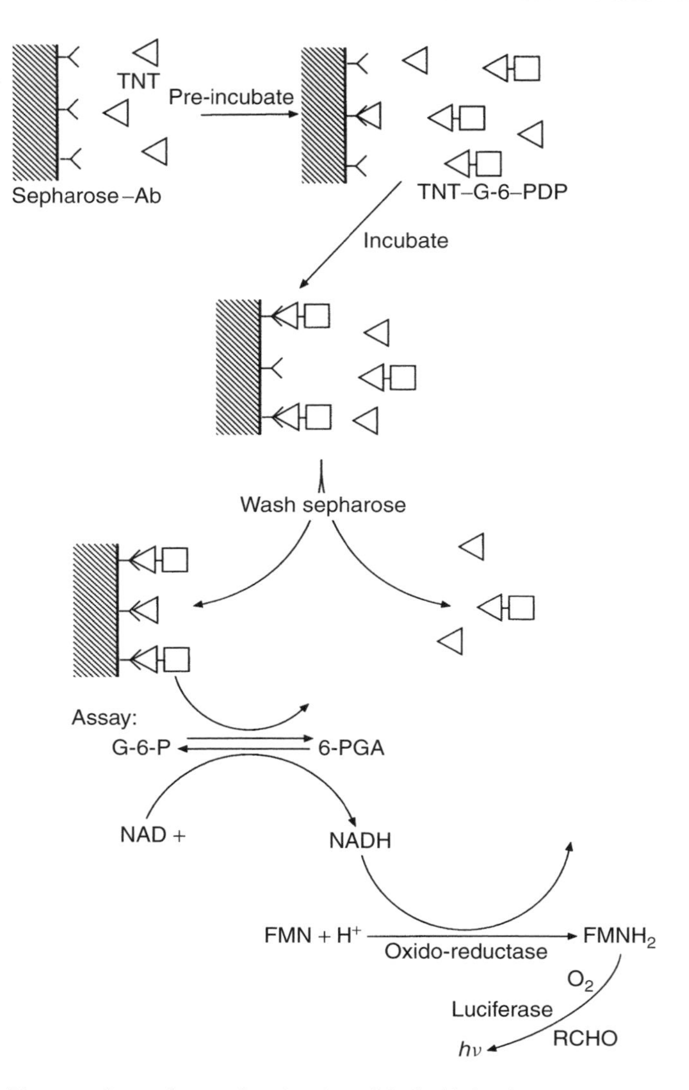

Figure 8.6 The reaction scheme for the (amplified) bioluminescent assay of trinitrotoluene. From Hall, E. A. H., *Biosensors*, Copyright 1990. © John Wiley & Sons Limited. Reproduced with permission.

The overall reaction can be summarized as follows:

$$\text{S–Ab} + \text{TNT} = \text{TNT:S–Ab} \longrightarrow + \text{TNT–glucose-6-phosphate–PDP}$$
$$\longrightarrow \text{wash, separate} \longrightarrow \text{assay}$$

$$\text{G-6-P (glucose-6-phosphate)} + \text{NAD}^+ \xrightleftharpoons{\text{TNT-G-6-PDH}}$$

$$\text{G-6-L-6-P (glucono-6}'\text{-lactone-6-phosphate)} + \text{NADH}$$

$$\text{NADH} + \text{FMN} + \text{H}^+ \longrightarrow \text{FMNH}_2 + \text{NAD}^+$$

$$\text{FMNH}_2 + \text{O}_2 + \text{RCHO} + \text{luciferase} \longrightarrow h\nu$$

SAQ 8.4

How could this method be modified to determine other explosive materials?

SAQ 8.5

What is the main source of error in this assay and how can it be eliminated?

8.5 Determination of Flavanols in Beers

8.5.1 Background

A plant-tissue biosensor which uses banana is very easy to construct. This was originally developed for the determination of dopamine, a brain chemical. In fact, the enzymes it contains, i.e. polyphenol oxidases, are effective with any catechol-type compounds, e.g. 1,2-dihydroxybenzene. The polyphenol oxidases catalyse the oxidation of catechols (by ambient oxygen) into *o*-quinones, as shown in Figure 8.7. The quinones can readily be detected by electrochemical reduction at a carbon paste electrode. We can combine the enzymatic stage with the electrochemical stage by mixing the banana with the carbon paste. This can then be used to determine certain catechol-type substances (catechins), also known as *flavanols*, which are found in beers and wines.

When this technique was first developed for use with beers, it was decided to use the more active, commercially available tyrosinase, as its performance characteristics were more reproducible from sample to sample. A screen-printed

Catechol $\xrightarrow[\text{Enzyme}]{\text{O}_2/\text{PPO}}$ *o*-quinone $\xrightarrow[\text{+ 2H}^+ \text{ Electrode}]{+ 2e^-}$ catechol

Figure 8.7 The enzymatic catalysis of catechols by polyphenol oxidase. From Eggins, B. R., *Biosensors: An Introduction*, Copyright 1996. © John Wiley & Sons Limited. Reproduced with permission.

biosensor was made and tested successfully. The main drawback was that beers do contain tyrosine, so the biosensor measures the combined total of catechins and tyrosine, whereas the *p*-dimethylaminocinnamaldehyde (DAC) colorimetric test used in the brewery industry just determines the catechins. Interestingly, the polyphenol oxidases in the original banana-based biosensor do not respond to tyrosine, so this device appears to be more selective.

8.5.2 Aim

8.5.2.1 Special Criteria for this Application

To monitor, both on-line and off-line, the flavanol content of beer during its production – excessive flavanols can cause cloudiness. Flavanols are antioxidants and help minimize heart disease.

8.5.2.2 Decide on the Selective Element

We will use ‘polyphenol oxidase’, which is found in many fruits and vegitables, such as bananas, apples, potatoes, mushrooms, etc. This oxidase is available from Sigma as ‘tyrosinase’, obtained from mushrooms.

8.5.2.3 Select the Transducer

An amperometric potentiostat (in the flow-injection analysis (FIA) mode) will be suitable for this application.

8.5.2.4 Decide on the Method of Immobilization

Entrapment in modified polypyrrole, followed by fabrication into screen-printed disposable electrode strips.

8.5.2.5 The Performance Factors Required

Calibration with (+)-catechin gives a range of 2.5 to 440 μM sensitivity, plus a response time of 48 s or 75 samples per hour, and a lifetime in which the current decreased by 7% over 6 h.

8.5.2.6 The Read-Out Mode

A BAS LC-4C amperometric detector was used, together with a Lloyds X-Y-t recorder.

8.5.2.7 Construction of the Sensor

Initial experiments used a carbon paste electrode mixed with either plant tissue or tyrosinase, while later work used screen-printed electrodes containing this enzyme. The latter was dissolved in amphiphilic substituted-pyrrole. 10 μl of sample were deposited on the surface of a screen-printed carbon electrode, and

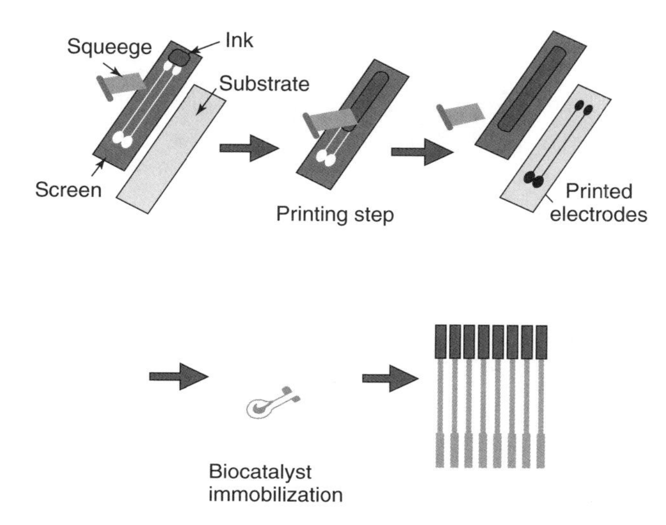

Figure 8.8 Schematic of the fabrication process of a screen-printed electrode containing an enzyme–amphiphilic substituted-pyrrole on the electrode surface. From Cummings, E. A., *PhD Thesis*, University of Ulster, UK, 2000.

electropolymerization was then carried out at 0.9 V in 0.1 M $LiClO_4$ at pH 7.0 for 20 min. The modified electrodes were washed and stored in a refrigerator. The fabrication process is illustrated schematically in Figure 8.8.

8.5.2.8 Testing of the Sensor

The flavanols in beer consists of a mixture of (+)-catechin and (−)-epicatechin, together with various dimers and trimers. These can be separated by using HPLC. However, the polyphenol oxidase responds to all of these compounds, albeit at different sensitivities. Thus, this device can be used to detect the total flavanol content by using either catechol or (−)-epicatechin as the standard. The latter had a sensitivity of 6.9 mA M^{-1}, with a precision of ±5%.

Comparison was made with a colorimetric method, using *p*-dimethylamino-cinnamaldehyde (DAC) (see Figure 8.9). The correlation obtained was between 0.96 and 0.99.

SAQ 8.6

What is the mediator in the above reaction?

Summary

This chapter presents the application of a number of sensor techniques to five different analyses.

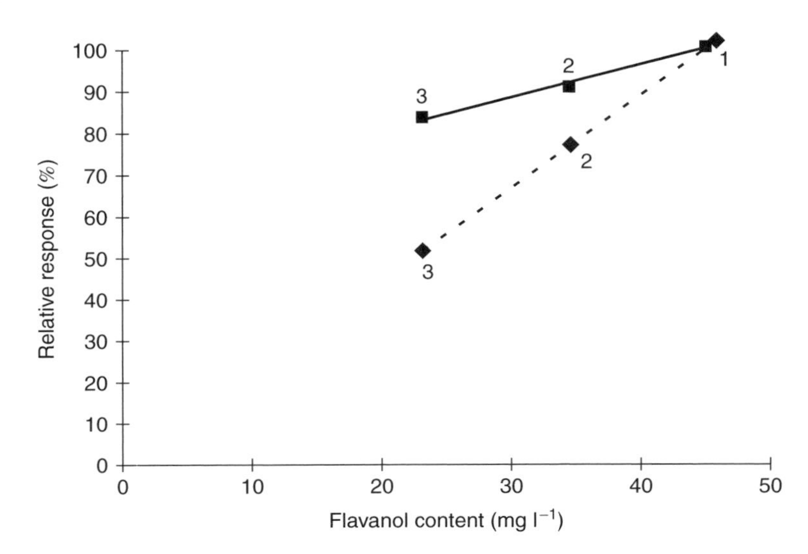

Figure 8.9 Comparative test results obtained for the analysis of flavanols in beers: (———) enzymatic biosensor device; (- - - -) colorimetric method, using *p*-dimethylaminocinnamaldehyde (DAC). From Cummings, E. A., *PhD Thesis*, University of Ulster, UK, 2000.

In the first, we see the structure and performance of the Medisense 'ExacTech' biosensor for the determination of glucose in blood. In the next application, low concentrations of copper(I) in water are analysed by complexation and differential pulse polarography. The third example outlines the developments towards a 'laboratory on a chip', where the use of arrays of sensors to determine mixtures of analytes is discussed. This is applied specifically to the employment of ISFET sensors for the simultaneous determination of four metal ions in blood. The fourth example describes how picomole amounts of explosive materials can be determined by the combination of an immunoassay linked via enzyme processes to a bioluminescent assay. The final application presents the use of a broad-spectrum enzyme available in plant tissues for the analysis of flavanols in beers, employing an amperometric technique.

Further Reading

Section 8.1

Cass, A. E. G., Davis, G., Francis, G. D., Hill, H. A. O., Aston, W. J., Higgins, I. J., Plotkin, E. V., Scott, L. D. L. and Turner, A. P. F., 'Ferrocene-mediated enzyme electrode for amperometric determination of glucose', *Anal. Chem.*, **56**, 6567–6571 (1984).

Raaskin, P., Strowig, S., Kilo, Ch., Dudley, J. D. and Ellis, B., 'Accuracy and precision of the ExacTech blood glucose system', in *ExacTech Pen and Companion: A New Generation of Blood Glucose Sensors* (Leaflet), MediSense, Birmingham, UK, 1987.

Section 8.2

Prabhu, S. V., Balwin, R. P., and Kryger, L., 'Chemical preconcentration and determination of copper at a chemically modified carbon paste electrode containing 2,9-dimethyl-1,10-phenanthroline', *Anal. Chem.*, **59**, 1074–1078 (1987).

Section 8.3

Forster, R. J., 'Miniaturised chemical sensors', in *Chemical and Biological Sensors*, Diamond, D. (Ed.), Wiley, New York, 1998, pp. 235–262.

Sibbald, A., Covington, A. K. and Carter, R. F., 'A quadruple function ISFET for clinical applications', *Med. Biol. Eng. Comput.*, **23**, 329–338 (1985).

Section 8.4

Wannlund, J., Egghart, H. and DeLuca, M., 'Bioluminescent immunoassays: A model system for detection of compounds at the attomole level', in *Luminescent Assays: Perspectives in Endocrinology and Clinical Chemistry*, Vol. 1, Serio, M. and Pazzagli, M. I. (Eds), Raven Press, New York, 1982, pp. 125–128.

Section 8.5

Cummings, E. A., 'A study of amperometric biosensors for the detection of phenolic compounds', *PhD Thesis*, University of Ulster, UK, 2000.

Cummings, E. A., Mailley, P., Liquette-Mailley, S., Eggins, B. R., McAdams, E. T. and McFadden, S., 'Amperometric carbon paste biosensor based on plant tissue for the determination of total flavanols content in beers', *Analyst*, **123**, 1975–1980 (1998).

Eggins, B. R., Hickey, C., Toft, S. A. and Zhou, D. M., 'Determination of flavanols in beers with tissue biosensors', *Anal. Chim. Acta*, **347**, 281–288 (1997).

Responses to Self-Assessment Questions

Chapter 1

Response 1.1

A labelled schematic of a chemical sensor is shown in Figure SAQ 1.1 below.

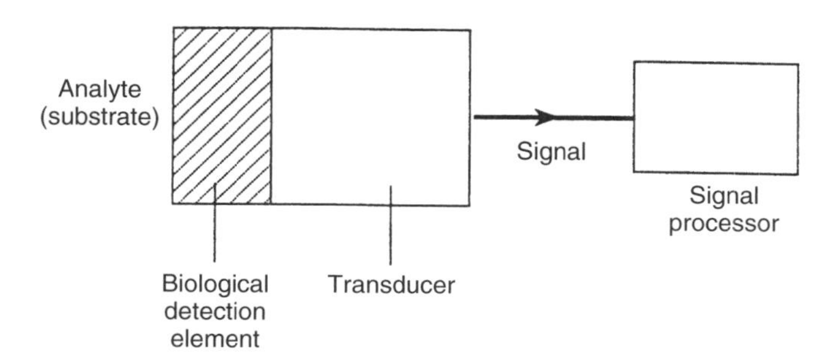

Figure SAQ 1.1 Schematic layout of a (bio)sensor. From Eggins, B. R., *Biosensors: An Introduction*, Copyright 1996. © John Wiley & Sons Limited. Reproduced with permission.

Response 1.2

We can divide sensors into three types, namely (a) *physical sensors*, for measuring distance, mass, temperature, pressure, etc. (which will not concern us here), (b) *chemical sensors*, which measure chemical substances by chemical or physical responses, and (c) *biosensors*, which measure chemical substances by using a biological sensing element. All of these devices have to be connected to a transducer of some sort, so that a visibly observable response occurs. Chemical sensors and biosensors are generally concerned with sensing and measuring particular chemicals which may or may not be biological in nature themselves.

Response 1.3

- All sensors must have a selective element that distinguishes one analyte from another.
- They must have a sensitive element that responds quantitatively to the amount of analyte.
- They must have a transducer that converts the sensor response into an observable form.
- They must have a means of connecting these elements together (immobilization).

Response 1.4

(a) A glucose sensor for diabetes, or a urease monitor for urea.
(b) An alcohol sensor for use in the brewing industry.
(c) A phenolics monitor for pollutants.

Chapter 2

Response 2.1

We have to be able to make a complete circuit to allow electrical measurements to be made. One of the electrodes is potentially a source of electrons, while the other acts as a 'sink' (see Figure SAQ 2.1 below).

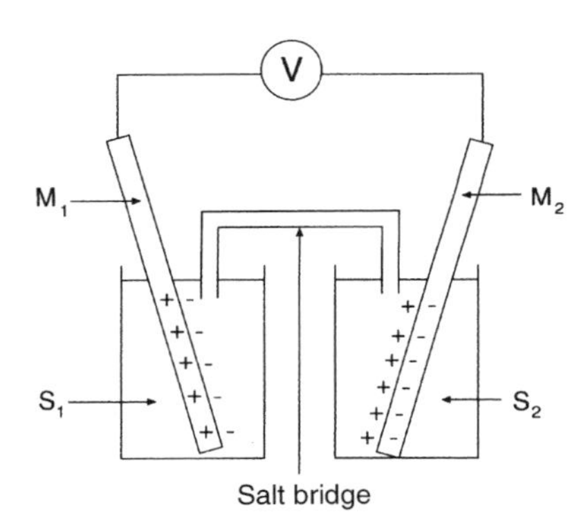

Figure SAQ 2.1 Two half-cell electrodes combined, making a complete cell. From Eggins, B. R., *Biosensors: An Introduction*, Copyright 1996. © John Wiley & Sons Limited. Reproduced with permission.

Response 2.2

The cell reaction is as follows:

$$Cu^{2+} + 2e^- \longrightarrow Cu \quad (1)$$

$$Fe^{3+} + e^- \longrightarrow Fe^{2+} \quad (2)$$

By subtracting '2×' equation (2) from equation (1), we obtain:

$$2Fe^{2+} + Cu^{2+} = 2Fe^{3+} + Cu \quad (3)$$

Response 2.3

A reference electrode must have the following properties:

- It should be easy to make
- It must provide a stable potential
- It must be polarizable (ideally)
- It should have a low temperature coefficient

Response 2.4

Consider two identical half-cells in a concentration cell on either side of an ion-selective membrane, differing in the concentrations of the ions. For each half-cell, we can write a Nernst equation, as follows:

$$E_1 = E^0 + RT/nF \ln c_1 \quad (1)$$

$$E_2 = E^0 + RT/nF \ln c_2 \quad (2)$$

By subtracting equation (2) from equation (1), we obtain:

$$E = E_1 - E_2 = -S \log c_2 (= 0) + S \log c_1$$

which gives:

$$E = K + S \log c_1$$

where S is the *slope* $= 2.303RT/nF = 0.059/n$, and K is the *intercept* $= -S \log c_1$ (c_1 is the standard concentration).

Response 2.5

$$E = E^0 + (0.059/2) \log (a_{Zn})/1$$

$$\therefore \; -0.789 = -0.76 + 0.029 \log (a_{Zn})$$

$$\therefore \; \log (a_{Zn}) = [(-0.789 - (-0.76)]/0.059 = -0.029/0.029 = -1.0$$

and therefore:

$$a_{Zn} = 0.1 \text{ M}$$

Response 2.6

The emf (potential) of a cell is proportional to the logarithm of the analyte concentration.

Response 2.7

By plotting the data shown in the table as $E = S \log [\text{Ca}] + K$, we obtain values for the slope S of 26.557 and K (the intercept) of 104.16.
Hence, $E = 26.557 \log [\text{C}] + 104.16$.
If $E_{(S)} = 33$, $\log [\text{S}] = -2.677$, and therefore $[\text{S}] = 2.09 \times 10^{-3}$ M.

Response 2.8

(a) The diffusion current is the current caused by a concentration gradient from a high- to a low-concentration region. This current is proportional to the slope of the concentration gradient and to the diffusion coefficient of the diffusing species, i.e. $I_d = \text{D}dc/dx$.
(b) The diffusion current is directly proportional to the concentration of the analyte.

Response 2.9

(a) The concentration of the oxidized species (Ox) at the electrode surface is not zero at the peak potential and only becomes zero at potentials considerably past $E_{p(c)}$. A similar situation exists for the reduced species (R) at the anodic peak potential. This causes a small shift of each peak, resulting in a peak separation of $0.058/n$ V.
(b) The standard redox potential is the average of the anodic and cathodic peak potentials for a reversible process: $E^0 = 1/2(E_{p(a)} + E_{p(c)})$.

Response 2.10

(a) The current is directly proportional to the concentration.
(b) The current is directly proportional to the electrode surface area.
(c) The current is proportional to one over the square root of the time ($\propto 1/t^{1/2}$).
(d) The current is proportional to the square root of the sweep rate ($\propto v^{1/2}$).

Response 2.11

In a kinetic wave, there is a follow-up reaction to generate a new chemical species, whereas with a catalytic wave the original reactant is regenerated by the follow-up chemical reaction. This results in an enhanced current for the catalytic wave.

Response 2.12

A DC potential will cause positive ions to accumulate at the cathode and negative ions to accumulate at the anode. The cell thus becomes polarized and the current will slowly fall towards zero. However, with an alternating current the ions effectively oscillate with each current reversal and so a quasi-equilibrium situation is reached, with the alternating current being constant, and can thus be measured.

Response 2.13

As the gate voltage is increased, there is a depletion of holes (for p-type silicon) until the concentrations of electrons and holes near the interface become equal, with the Fermi level now mid-way between the valence band and the conduction band. When the gate voltage is increased further, there is an excess of electrons at the interface, thus resulting in an inversion of the p-type silicon to become n-type silicon (see Figure 2.22 (d) in the text).

Response 2.14

Carbon paste is an ideal material with which biomaterials can be mixed, with banana being one of the easiest materials to handle in this way. It contains polyphenoloxidase enzymes, which are selective for the oxidation of a range of polyphenols, including dopamine and catechins, found in beers and wines. The mixture of carbon paste and banana can be screen-printed to form thick-film electrodes.

Response 2.15

- They can operate with very low currents
- They have very low double-layer capacitances, and hence low capacitive currents
- They have minimal IR values
- They can operate at high sweep rates
- They can be used in a two-electrode mode
- They can be inserted into the brain (for example)

Response 2.16

$$A = \log(I/I_0) = \varepsilon C l$$

where A is the absorbance, I the intensity of the transmitted light, I_0 the intensity of the incident light, ε the extinction coefficient, C the concentration of analyte, and l the pathlength of the cell.

Response 2.17

Using the Beer-Lambert law:

$$A = \varepsilon Cl$$

and so:

$$0.85 = 1000 \times C \times l$$

which gives:

$$C = 0.85/1000 = 8.5 \times 10^{-4}\ \text{M}$$

Response 2.18

Chemiluminescence is produced in a chemical reaction in which light is produced without heat. The best-known example of this phenomenon is the reaction of luminol with oxygen in alkaline solution.

Response 2.19

Bioluminescence may be produced from the luciferins found in bacteria, or in glowworms or fireflies.

Response 2.20

When the angle of incidence of a light beam on a second medium becomes greater than θ_c, the critical angle, it does not penetrate the medium to be refracted, but is totally reflected. Figure SAQ 2.20, shown below, illustrates this behaviour.

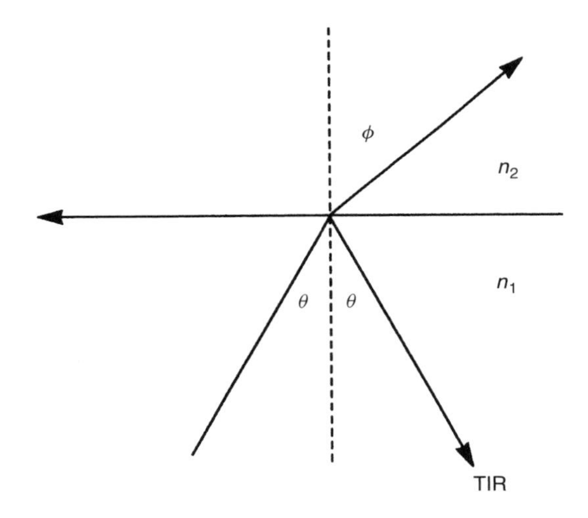

Figure SAQ 2.20 Illustration of total internal reflection. From Eggins, B. R., *Biosensors: An Introduction*, Copyright 1996. © John Wiley & Sons Limited. Reproduced with permission.

Response 2.21

The light beam is totally internally reflected along the internal surface of the fibres and so is contained by them (see Figure SAQ 2.21(a) below).

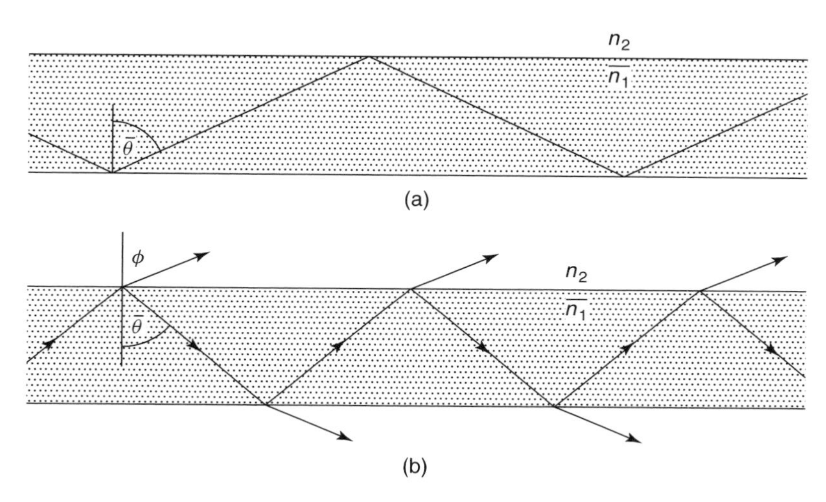

Figure SAQ 2.21 Illustration of total internal reflection in a single-core optical fibre, showing the trajectories for (a) bound and (b) refracting rays.

Response 2.22

The most inexpensive light sources are light-emitting diodes (LEDs) which generally operate at a narrower emission bandwidth ($\approx$50 nm) when compared with incandescent lamps. As well as their small size, they are of low cost and can be easily connected to optical fibres. High-intensity blue LEDs are now available for fluorescence studies.

Chapter 3

Response 3.1

The correct answer is (a), i.e. to ensure that all of the solutions have the same total ionic strength.

Response 3.2

They can act as ion-exchange materials, or can entrap enzymes or antibodies in the polymer matrix.

Response 3.3

The counter-ion can greatly modify the relative ion-exchange properties of the (conducting) polymer. By altering the counter-ion, the polymer can change from a cation exchanger into an anion exchanger.

Response 3.4

An ionomer film coated on an electrode can extract and pre-concentrate large electroactive ions, such as methyl viologen, which can then be detected by voltammetry. This process is known as ion-exchange voltammetry.

Response 3.5

A redox group, such as a quinone, can be covalently bonded to a monomer, such as 4-vinylpyridine, which is then polymerized on the surface of the electrode.

Response 3.6

They have different distribution coefficients for different ions. However, this selectivity is fairly limited.

Response 3.7

From the Michaelis–Menton equation:

$$v = \frac{k_2[E_0][S]}{K_M + [S]}$$

We can see that when $K_M \gg [S]$ the rate, v, is linearly proportional to [S], i.e. $v = k_2[E_0][S]/K_M$. However, as [S] increases and becomes significant when compared to K_M, this equation becomes non-linear. Eventually, when $K_M \ll [S]$, the relationship becomes independent of [S], and $v = k_2[E_0]$, which is the maximum value of $v\,(= v_{max})$.

Response 3.8

Some enzymes are catalysts for a range of similar reactions. For example, polyphenoloxidases will catalyse the oxidation of a wide range of polyphenols and related substances. The actual activity varies depending on the source of the enzyme. Sometimes, even 'purified' enzymes may consist of a mixture of related enzymes.

Response 3.9

Often, an assay is required to measure the total amount of analyte, whereas the antibody is selective for one isomer only.

Response 3.10

With the rapid expansion in knowledge of the structure and role of genes in biological systems, and advances in the technology of how to manipulate them, there are a increasingly larger number of applications for nucleic acids in sensor devices.

Response 3.11

An antibody binds closely with its complementary antigen. Indeed, an antigen can be induced to create the corresponding antibody. Receptors, however, are like '*messengers*' which transmit signals between different parts of a biological system. The signal may be chemical in form, so the receptor can therefore respond to a particular chemical (or group of chemicals).

Response 3.12

Covalent bonding > cross-linking > microencapsulation > encapsulation > adsorption.

Response 3.13

In adsorption, the forces connecting the absorbent and absorbate are very weak, consisting mainly of van der Waals forces and also possibly some hydrogen bonding. These attachments are not very stable or permanent, and so the lifetimes of sensors made in this way will be limited.

Response 3.14

Despite the response to SAQ 3.13 given above, antibodies are often strongly adsorbed on to a surface, particularly when associated with the corresponding antigen. Modification of the surface may enhance the strength of this binding.

Response 3.15

Bacteria are normally immobilized on membranes, i.e. by using the microencapsulation technique.

Chapter 4

Response 4.1

Specific means that the sensor responds to one analyte only, whereas *selective* means that it responds to a range of analytes, but that the response to the one of interest is much greater than to the others.

Response 4.2

The maximum allowable error is 10% on 10^{-6} M, i.e. 10^{-7} M.

The selectivity coefficient tells us that 1 M OH^- will give the same response as 10^{-1} M F^-, and therefore the concentration of OH^- which will give a response similar to 10^{-6} M F^- is 10^{-8} M, i.e. pOH = 8.

In water, this corresponds to a pH of (14 − 8) = pH 6. Therefore, in order to guarantee minimum interference from hydroxide ions, the pH of the test solution must be adjusted to less than ph 6 (pH 5.5 is usually used). At this pH, the worst interference effect would be 10% at 10^{-6} M fluoride (which is about the limit of detection for the ion-selective electrode).

Response 4.3

These include the following:

- Temperature of storage
- pH of storage
- Basic stability of the enzyme
- Environment of the enzyme, e.g. a pure enzyme versus a tissue-based or micro-organism-based material
- Special stabilizing additives, such as DEAE–dextran plus lacticol

Response 4.4

Accuracy describes the proximity to the true value, which is affected by systematic errors, while *precision* (which is the same as reproducibility) describes the magnitudes of the random errors.

Response 4.5

Because potentiometric sensors operate on a logarithmic scale, the linear range may be much larger than those of amperometric sensors. For example, a pH electrode has a range of twelve powers of ten or 12 pH units. The effect of this may be to make the precision rather less than that found for an amperometric sensor.

Response 4.6

(a) A glass membrane.
(b) A lanthanum fluoride crystal.
(c) Glucose oxidase enzyme.
(d) An antibody to oestradiol.
(e) Receptors or antibodies.

Chapter 5

Response 5.1

One does not need to know the cell emf. For the test sample on its own:

(a) Assume that the slope is Nernstian.
(b) Calculate the concentration for each addition, $C = V \times 0.1/(50 + V)$.
(c) Subtract the blank value from each emf reading.
(d) Calculate $10^{E/S}$ for each addition.
(e) Plot $10^{E/S}$ versus C.

Volume added (cm^3)	C (M)	C ($M \times 10^{-3}$)	E (mV)	$10^{E/S}$
1.0	$1 \times 0.1/51$	1.96	$99.8 - 70 = 29.8$	3.193
2.0	$2 \times 0.1/52$	3.85	$102.5 - 70 = 32.5$	3.547
3.0	$3 \times 01/53$	5.66	$104.6 - 70 = 34.6$	3.850
4.0	$4 \times 0.1/54$	7.41	$106.3 - 70 = 36.3$	4.114
5.0	$5 \times 0.1/55$	9.091	$107.9 - 70 = 37.9$	4.378

The negative intercept on the concentration axis gives the value for the unknown. NB – as this is a Gran plot in which we are plotting the inverse logarithm of the emf against the concentration, the method gives the actual concentration (not the logarithm of concentration). We obtain a result where $C_u = 0.0175$ M.

Response 5.2

Set up two equations for each measurement, assuming Nernstian behaviour, with $S = 59$ mV:

$$98 = K + 59 \log C_u \quad (1)$$

$$73 = K + 59 \log (C_u \times 50 + 100 \times 2)/52 \quad (2)$$

Subtract equation (2) from equation (2) to eliminate K:

$$98 - 73 = 25 = 59 \log C_u/[(C_u \times 50 + 100 \times 2)/52]$$

Then take the antilogarithms of both sides, and solve for C_u:

$$C_u = 100[10^{25/59}(1 + 50/2) - 50/2]$$

Thus, $C_u = 100 \times 2.653 \times 1 = 265.3$ ppm. Therefore, in the original tea sample, the concentration of fluoride ions, $C_{F^-} = 530.6$ ppm.

Response 5.3

Potentiometric gas sensors are based largely on the acid/base properties of the gas. The gas-sensing membrane may have some selectivity. However, the main effect is through the buffering internal electrolyte solution (see Figure 5.4). Further details are given in Table 5.2.

Response 5.4

Urea is easily determined with a potentiometric biosensor based on urease (obtained from jack bean meal). This catalyses the hydrolysis of urea to ammonia and carbon dioxide:

$$CO(NH_2)_2 + H_2O \longrightarrow 2NH_3 + CO_2$$

The pH can be adjusted to allow determination via several different electrode systems, including NH_3, $NH_4{}^+$, CO_2 or pH. The latter can be made simply from a glass pH electrode by immobilizing the urease in a gel, although its performance characteristics are poor. Table 4.3 shows a comparison of the different types from which it appears that the cation-based biosensor is the most versatile. An even simpler version can be made by using a platinum electrode coated with urease immobilized in polypyrrole, although its performance is variable. An amperometric sensor has been described in which a hydrazine electrode is used:

$$N_2H_4 + 4OH^- \longrightarrow N_2 + 4H_2O + 4e^-$$

The current is very pH-dependent, so it will respond to changes in pH from the urease reaction.

Response 5.5

A *mediator* is an electron-transfer agent, usually a reversible oxidizing agent which will transfer electrons between an enzyme and an electrode.

Response 5.6

(a) The ambient level of oxygen must be controlled and constant.
(b) Oxygen has a fairly high reduction potential (−0.7 V) at which other materials might interfere.
(c) The redox behaviour of oxygen is not very reversible.

Response 5.7

Iron(III) ions are readily hydrolysed to the insoluble $Fe(OH)_3$ ($+3H^+$). In addition, the redox potential of Fe(III/II) is rather high (0.53 V at pH 7).

Response 5.8

- Natural – cytochrome c_3, vitamin K_2, NAD^+/NADH
- Artificial – ferrocenes, Methylene Blue, methyl viologen

See Table 5.6 for further examples.

Response 5.9

(a) The electrochemical reduction of NAD^+ does not give NADH, but goes to a dimer. Although NADH can be oxidized directly to NAD^+, this can only be achieved at a substantial over-potential.
(b) The NAD^+/NADH couple may be linked through another mediating enzyme, such as diaphorase (lipomide dehydrogenase). Another alternative is to use a conducting salt electrode, such as $NMP^+/TCNQ^-$.
(c) Examples include lactate to pyruvate, cholesterol to cholestenone and ethanol to acetaldehyde (see Table 5.7 for other possibilities).

Response 5.10

A number of amperometric biosensors can be constructed for lactate determination, mostly involving lactate dehydrogenase, although other enzymes such as lactate oxidase or lactate monooxidase may be used (see Figure SAQ 5.10). A variety of mediators have been used with these systems, including NAD^+/NADH and $Fe(CN)_6{}^{3-}$. Sometimes, two enzymes can be coupled together to produce an amplification effect which can improves the sensitivity by up to 4000 times.

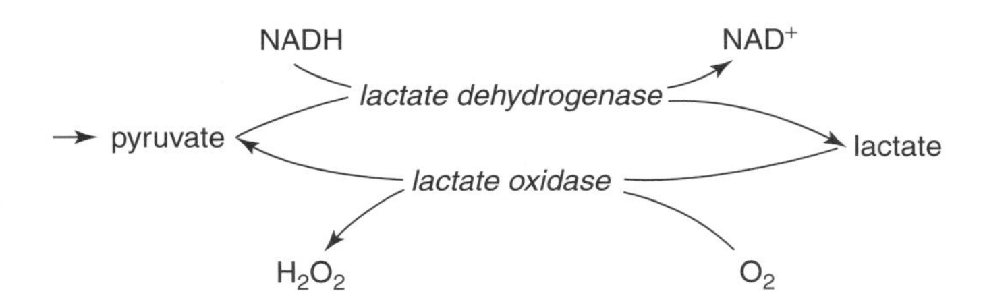

Figure SAQ 5.10 Determination of lactate via the use of oxidase or dehydrogenase biosensor systems.

Response 5.11

- IGFET – this is the insulated-gate field-effect transistor (FET), which is the basic design.
- CHEMFET – this is an FET where a chemically active substance modifies the gate.

- ISFET – this is an FET where an ion-selective material, such as an ionophore, replaces the gate metal.
- ENFET – this is an FET where an enzyme replaces the gate metal.

Chapter 6

Response 6.1

Extrinsic fibre sensors are those in which the function of the optical fibres is to convey the light to and from the sensing location, which may or may not employ chemical reagents, whereas in an *intrinsic* sensor the optical fibre plays an active role, so that the transmitted light may be modulated by chemically induced interactions in the fibre core or in the fibre cladding. For example, the fibre cladding may be chemically sensitive and interact with the light via the evanescent field.

Response 6.2

A pH optode has been commercially developed for analysis of the pH of blood. The pH range involved is very narrow, i.e. pH 7.0–7.5. This is much narrower than the usual electrochemical pH range, but may be more accurate over this particularly crucial range. A signal-to-noise ratio (S/N) of 2000 to 4000 has been achieved.

Response 6.3

(i) The fluorescence signal increases with analyte concentration and is measured against a zero background light level.
(ii) The light is emitted by an analyte or reagent molecule and contains more information about these species. One can measure, for example, the fluorescence intensity, decay time, polarization anisotropy and energy transfer.

Response 6.4

When a light ray is totally internally reflected in an optical fibre within a certain range of angles, a fraction of the radiation extends a short distance from the light-propagating medium into the medium of lower refractive index. This non-propagating electromagnetic radiation is called the *evanescent* wave.

Response 6.5

For plane-polarized light incident on an interface between two phases of different refractive indexes, the reflection, R_p, is zero for the Brewster angle, i.e. when $\theta = \tan^{-1}(n_1/n_2)$, where n_1 and n_2 are the refractive indexes of the two phases and θ is the angle of incidence. This may be modified for a transparent incident phase.

Response 6.6

A surface plasmon is a collective oscillation of free electrons in a metal film.

Response 6.7

(i) It functions as a highly sensitive refractometric process with a selectivity imparted by the specific chemical–biochemical material coated on the sensing side of a metal film. For example, antibody–antigen reactions on the surface of a metal change the dielectric constant and result in a shift in the resonance angle.
(ii) It can provide information about the layer structure and thickness without the use of more complex instrumentation.
(iii) Optical excitation of the immobilized layer is possible via the surface plasmon wave with considerable enhancement.
(iv) It is a highly sensitive means of providing information about the molecular layer structure. SPR-enhanced Raman spectra can be recorded for organic films as thin as a monolayer.
(v) There can be real-time monitoring of binding kinetics.

Response 6.8

- Rayleigh scattering – resulting from very small particles
- Rayleigh–Gans–Debye scattering
- Mie scattering

Chapter 7

Response 7.1

(a) Certain isotropic crystals will vibrate when an electrical potential is applied to them; conversely, a vibrating crystal will generate an electrical potential.
(b) Quartz, tourmaline, barium titanate, lead titanates, and polymers such as poly(vinylidene fluoride) all display piezo-electric properties.

Response 7.2

The change in resonant frequency (Δf) resulting from the adsorption of an analyte on the surface of a piezo-electric material can be measured with high sensitivity (500–2500 Hz g^{-1}), thus resulting in sensor devices with picogram (pg) detection limits.

Response 7.3

A selective material, which will interact selectively with the analyte, e.g. antibodies or enzymes, needs to be adsorbed on the surface of the crystal.

Response 7.4

Detection limits as low as 10^{-12} g (1 pg (picogram)) can be obtained when using this type of sensor.

Response 7.5

A thermistor is a very sensitive device for measuring changes in temperature, particularly in biochemical (enzymatic) reactions. A typical thermistor detector contains a resistance thermometer consisting of a 16 kΩ resistance with a coefficient of −3.9% per degree. The resistance is measured against a reference source by using a DC Wheatstone bridge amplified to give an output of up to 100 mV per 0.001 degrees.

Response 7.6

This consists of a platinum wire resistance thermometer/heating wire embedded in a ceramic bead (1 cm in size), and is used for the detection of combustible gases in air. The bead is coated with finely divided palladium in a thorium oxide matrix to catalyse combustion of the gases. The pellister may be made resistant to catalyst 'poisons', such as alkyllead, organosulfur and organophosphorus compounds, by surrounding the resistance wire with alumina containing a large amount of the catalyst.

Response 7.7

- Calorimetric – measurement of the heat evolved in a biochemical reaction.
- Catalytic – catalytic oxidation of flammable gases and measurement of the heat evolved.
- Thermal conductivity – here, a heated filament is placed in the path of a gas. Its resistance varies with temperature, while the temperature varies with the thermal conductivity of the surrounding gas.

Chapter 8

Response 8.1

We could use an implanted subcutaneous sensor, which could be left in place for at least several days, and preferably weeks, at a time. The biosensor would need to be biocompatible, would have to be small so as to cause minimal damage during insertion. It should have a linear response of 0–20 mM (±1 mM), with a response time better than 10 min. External calibration should be possible within ± 10% over 24 h. One might imagine a more complex, permanently implanted unit with a radio device to transmit data, but this would be vastly more expensive. We could also conceive of an iontophoretic sensor that could make external measurements of electrical signals through the skin without actual implantation.

Response 8.2

The approach differs from the conventional cathodic stripping analysis in that the concentration step is not electrochemical, but is effected by the chemical extraction of the analyte to form a complex with the 2,9-DMP, without any change in the oxidation state. However, the actual analytical step is by cathodic stripping – reduction of the Cu(I) complex to Cu. Thus, it is, in principle, a cathodic stripping analysis.

Response 8.3

Photometric sensors for all of these ions are available, although the selectivity between sodium and potassium is not particularly good. For H^+, Na^+ and K^+, fluorescent methods exist, while for Ca^{2+} a visible spectrophotometric method is available. Combining these into a single sensor device would be very difficult!

The following sensor systems have been used:

- pH – an 8-hydroxy-1,3,6-pyrenetrisulfonic acid (HPTS) fluorescent sensor, $\lambda_{ex} = 405/470$ and $\lambda_{em} = 520$ nm.
- Na^+ – the optically labelled fluor(8-anilino–1-naphthalene sulfonate) in an immobilized ionophore–Na^+ complex, in competition with a quenching cationic polyelectrolyte, poly(copper(II)-polyethyleneimine).
- K^+ – potassium sensors has been made based on a 2-hydroxy-1,3-xylyl-18-crown-5 (crown ether) with diazotised 4-nitroanailine immobilized at the end of an optical fibre. Although the detection limit is 0.5 mM, the selectivity against sodium (1/6) would be insufficient for clinical applications.
- Ca^{2+} – a technique has been described in which a calcium-selective ionophore (L) extracts the calcium into a PVC membrane where it interacts with a chromophoric pH indicator (CH^+) incorporated in the sensor membrane. To maintain electroneutrality, H^+ ions attached to the protonated basic indicator are exchanged and released into the sample:

$$Ca^{2+} + L + 2CH^+ \longrightarrow 2H^+ + CaL^{2+} + 2C$$

The indicator used shows deprotonation by a decrease in absorbance at 660 nm. It has been suggested that this approach could be developed for use with a wide variety of ions.

Response 8.4

An antigen corresponding to the particular explosive material would need to be substituted for the TNT–antibody. It is doubtful if an antibody could be found for all explosives in general, although a mixture of antibodies corresponding to every likely explosive material might work.

Response 8.5

Because of the very low levels of analyte concentration, there is always the danger of contamination, especially if the laboratory workers have been handling authentic samples of TNT for carrying out the determination of standards.

Such contamination can only be eliminated by scrupulous cleaning of all vessels, containers, etc., and by the use of (ultra) high-purity reagents. It is also preferable that a different person to the one who set up the system and ran the standards should be employed to carry out the actual analysis.

Response 8.6

Ambient oxygen is the mediator. Although the best way of carrying out the analytical measurement is by electrochemical reduction of the product quinones, following the uptake of oxygen (as in the original glucose biosensor) could be a possible approach. Banana-based biosensors for dopamine have been used in this way. An advantage when analysing a mixture of different polyphenols would be that the response to oxygen uptake would be independent of the nature of an individual polyphenol, whereas we have shown in our discussion that the diffusion currents for the reduction of the quinones from different catechins and catechol are quite different. This is due largely to the different diffusion coefficients resulting from the different sizes of the molecules.

Bibliography

Bard, A. J. and Faulkner, L. R., *Electrochemical Methods and Applications*, 2nd Edn, Wiley, New York, 2001.

Catterall, R. W., *Chemical Sensors*, Oxford University Press, Oxford, UK, 1997.

Diamond, D. (Ed.), *Principles of Chemical and Biological Sensors*, Wiley, New York, 1998.

Eggins, B. R., *Biosensors: An Introduction*, Wiley, Chichester, UK, 1996.

Eldham, P. G. and Wang, J. (Eds), *Biosensors and Chemical Sensors*, American Chemical Society, Washington, DC, USA, 1992.

Hall, E. A. H., *Biosensors*, Open University Press, Buckingham, UK, 1990.

Janata, J., *Principles of Chemical Sensors*, Plenum Press, New York, 1989.

Miller, J. C. and Miller, J. N., *Statistics for Analytical Chemistry*, 3rd Edn, Ellis Horwood, Chichester, UK, 1993.

Monk, P. M. S., *Fundamentals of Electroanalytical Chemistry*, AnTS Series, Wiley, Chichester, UK, 2001.

Scheller, F. and Schmid, R. D. (Eds), *Biosensors: Fundamentals, Technologies and Applications*, GBF Monographs, Vol. 17 (Proceedings of an International Seminar, May 12–14 1991, Bogensee/Brandenburg, Germany), VCH, Weinheim, Germany, 1992.

Turner, A. P. F., Karube, I. and Wilson, G. S. (Eds), *Biosensors: Fundamentals and Applications*, Oxford University Press, Oxford, UK, 1987.

Usher, M. J. and Keating, D. A., *Sensors and Transducers*, 2nd Edn, Macmillan, London, 1996.

Glossary of Terms

This section contains a glossary of terms, all of which are used in the text. It is not intended to be exhaustive, but to explain briefly those terms which often cause difficulties or may be confusing to the inexperienced reader.

Absorbance Represents the amount by which the intensity of light is diminished after passing through an analyte. It is given by $A = \log(I/I_0)$, where I_0 is the incident intensity and I the transmitted intensity.

Acoustic wave A sound wave produced by applying a radiofrequency to a piezo-electric crystal which produces a mechanical stress in the material, resulting in the formation of a wave in the audible range.

Activity The thermodynamic concentration perceived at an electrode.

Activity coefficient, γ The ratio of the activity a to the concentration C, i.e. $\gamma = a/C$.

Adsorption The attachment of material on to a surface by either chemical or physical means.

Amperometry A technique for the measurement of a concentration as a function of current.

Analyte The material to be determined quantitatively or qualitatively.

Analyte analogue A substance similar to the analyte that has inherent optical characteristics, such as fluorescence, which may be added to the analogue as a label. It can be used if the analyte shows no spectral changes on bonding to a reagent.

Anion A negatively charged ion.

Anode The electrode at which oxidation (removal of electrons) occurs.

Antibody A protein substance, produced by an organism, which can bind with an invading antigen and remove it from harm.

Antigen A toxic substance or organism that can be removed by binding to an antibody.

Attenuated total reflectance (ATR) A technique by which an absorbing material, placed in contact with a reflecting surface, causes attenuation of the internally reflected light.

Beer–Lambert law Describes the relationship between absorbance (A) and concentration (C) in photometry: $A = \varepsilon Cl$.

Bioassay An analytical method using biological sensing material.

Bioluminescence The emission of light by certain biological species such as the firefly, the glowworm and some forms of bacteria.

Biosensor A device incorporating a biological sensing element connected to a transducer.

Brewster angle The angle defined as $\theta = \tan^{-1}(n_1/n_2)$ when the reflectance is equal to zero.

Catalytic wave An electrochemical enhancement of (electrochemical) current by a catalytic reaction in which the original reactant is regenerated in a follow-up reaction.

Cathode The electrode at which reduction (addition of electrons) occurs.

Cation A negatively charged ion.

Chemical sensor A device that responds quantitatively to a particular analyte in a selective way through a chemical reaction.

Chemiluminescence The production of light from a chemical reaction 'in the cold' and in the absence of any exciting illumination.

Chronoamperometry The technique of measuring current as a function of time.

Covalent attachment Formation of covalent bonds between a biological material and a transducer.

Cross-linking A method for linking a transducer to a biological material using a bifunctional agent which forms chemical bonds with both.

Cyclic voltammetry The technique of measuring current as a function of potential in an experiment in which the potential is swept forwards and then back again.

Differential pulse voltammetry A form of voltammetry in which a liner potential sweep is applied to the working electrode, superimposed by a series of square-wave pulses. The current is measured in a differential mode, dI/dE, which is then plotted against the potential.

Diffusion Mass transport involving the movement of material from a high- to a low-concentration region.

Diffusion coefficient A measurement of the rate of diffusion, given as distance squared per unit of time.

Diffusion current The maximum current limited by diffusion.

Electrochemical quartz crystal microbalance (EQCM) A device used to measure very small mass changes on an electrode surface while immersed in an electrolyte solution (*see also* QCM).

Electrode An electronic conductor used to measure an electrode potential or a cell current.

Electrode potential The energy, expressed as a voltage, of a redox couple relative to an agreed standard.

Electrolyte An ionic salt in solution which conducts electricity.

Electromagnetic spectrum A display of the known range of electromagnetic waves, (usually) extending from γ-rays to radio waves, including visible light, UV and IR irradiation.

Entrapment Immobilization of a biomaterial within a three-dimensional polymer matrix.

Enzyme A large complex molecule, consisting largely of protein, and usually containing a prosthetic group, which often includes one or more metal atoms, and will catalyse a particular chemical reaction.

Enzyme electrode A biosensor in which the biological material is an enzyme and the transducer is an electrode.

Evanescent wave When a light ray is totally internally reflected in an optical fibre within a certain range of angles, a fraction of the radiation extends a short distance from the light-propagating medium into the medium of lower refractive index. This non-propagating electromagnetic radiation is called the evanescent wave.

Extrinsic sensors Those in which the function of the optical fibres is to convey the light to and from the sensing location, which may or may not employ chemical reagents.

Fluorescence An effect in which a substance absorbs light at one wavelength and re-emits it at a different wavelength.

Gran plot A multiple addition method applied to potentiometric methods, so that a plot of an anti-logarithmic function of the potential is made against concentration.

Half-cell potential The electrode potential resulting from a half-cell reaction.

Half-cell reaction A redox reaction at a single electrode.

Immobilization Attachment of a biological material (usually) to a transducer.

Immunoassay An assay in which the selective material is due to an antibody–antigen reaction.

Impedance AC resistance.

Intrinsic sensors Those in which the optical fibre plays an active role, so that the transmitted light may be modulated by chemically induced interactions in the fibre core or in the fibre cladding.

Ion A charged species.

Ion-exchange polymer A (polymeric) material which will exchange one ion for another.

Ion-selective electrode (ISE) A potentiometric chemical sensor for the measurement of ionic concentrations.

Light-emitting diode (LED) A diode used as a light source in photometric sensors.

Linear-sweep voltammetry A technique in which current is measured as a function of potential, where the latter is swept in a linear ramp.

Lineweaver–Burke plot A plot of 1/[S] versus $1/v$, used in enzyme kinetics.

Liquid junction potential A potential developed across a boundary between electrolytes which differ in concentration or chemical composition.

Luminescence This is a general term for techniques in which light is emitted, including fluorescence, phosphorescence, bioluminescence and chemiluminescence.

Mass transport Movement of material through a reaction medium.

Mediator A substance that will transfer electrons between a biological material (generally an enzyme) and an electrode. It is usually a redox material.

Michaelis–Menton equation The relationship between the concentration of a substrate and the rate of reaction in enzyme kinetics: $v = v_{max}[S]/(K_M + [S])$

Microelectrode (ultra-microelectrode) An electrode whose dimensions are in the micrometer (micron) range.

Microencapsulation Immobilization of a biological material in a membrane.

Migration Movement of charged species (particularly ions) under an electric field.

Mitochondria A multi-enzyme system found in cells outside of the nucleus.

Nernst equation The relationship between concentration (C) and potential (E) in a potentiometric measurement: $E = E^0 + RT/nF \ln C$.

Nernstian A system that obeys the Nernst equation.

Optode An optical chemical sensor.

Oxygen electrode A commonly used amperometric electrode for the analysis of oxygen, originally devised by L. C. Clark.

Quartz crystal microbalance (QCM) A device used for the measurement of very small changes of mass by using a piezo-electric crystal (*see also* EQCM).

Pellister A catalytic gas sensor based on a platinum wire heater and resistance thermometer.

Photometric Any analytical method involving the interaction of light with an analyte.

Piezo-electric crystal A crystal which vibrates mechanically on the application of an oscillating electrical potential and which will also produce an electrical potential if mechanically deformed.

Prosthetic group The part of an enzyme that is responsible for its catalytic activity.

Ramp A jargon term, meaning to smoothly increase at a constant rate of d(variable)/dt. A voltage ramp is therefore dE/dt.

Receptors These are proteins in the surface membranes of cells which, when bound to a ligand, initiate a particular physiological response in the cell through activation or blockage of a series of biochemical reactions.

Saturated-calomel electrode (SCE) A commonly used reference electrode based on calomel (mercury(II) chloride) and mercury in a saturated solution of potassium chloride.

Screen-printed electrode A miniature electrode, mass-produced by forcing a conducting paste through a template screen.

Selectivity The ability of a sensing element to discriminate between one chemical species and others.

Sensing element The active part of a sensor that responds to an analyte.

Standard electrode potential An electrode potential measured at standard temperature and pressure with all reactants and products present at unit activity.

Standard hydrogen electrode (SHE) The standard against which all other redox potentials are measured, based on the reaction $H^+ + e^- = 1/2H_2$ under standard conditions.

Substrate A substance to be analysed or reacting in an enzymic reaction (*see also* Analyte and Enzyme).

Surface plasmon resonance (SPR) An optical technique in which a plane-polarized incident field, which has an angle such that the photon momentum along the surface matches the plasmon frequency, couples to the electron plasma in the material (metal).

Thermistor A device used for measuring small changes in temperature in an enzymic or chemical reaction.

Transducer A device that converts an observed change (physical or chemical) into a measurable signal (usually electronic in nature) whose magnitude is proportional to the concentration of the analyte or analytes.

Voltammetry A technique in which current is measured against potential.

Voltammogram Plot of current against potential in voltammetry.

SI Units and Physical Constants

SI Units

The SI system of units is generally used throughout this book. It should be noted, however, that according to present practice, there are some exceptions to this, for example, wavenumber (cm^{-1}) and ionization energy (eV).

Base SI units and physical quantities

Quantity	Symbol	SI Unit	Symbol
length	l	metre	m
mass	m	kilogram	kg
time	t	second	s
electric current	I	ampere	A
thermodynamic temperature	T	kelvin	K
amount of substance	n	mole	mol
luminous intensity	I_v	candela	cd

Prefixes used for SI units

Factor	Prefix	Symbol
10^{21}	zetta	Z
10^{18}	exa	E
10^{15}	peta	P
10^{12}	tera	T
10^{9}	giga	G
10^{6}	mega	M
10^{3}	kilo	k

(*continued overleaf*)

Prefixes used for SI units *(continued)*

Factor	Prefix	Symbol
10^2	hecto	h
10	deca	da
10^{-1}	deci	d
10^{-2}	centi	c
10^{-3}	milli	m
10^{-6}	micro	μ
10^{-9}	nano	n
10^{-12}	pico	p
10^{-15}	femto	f
10^{-18}	atto	a
10^{-21}	zepto	z

Derived SI units with special names and symbols

Physical quantity	SI unit		Expression in terms of base or derived SI units
	Name	Symbol	
frequency	hertz	Hz	$1\ Hz = 1\ s^{-1}$
force	newton	N	$1\ N = 1\ kg\ m\ s^{-2}$
pressure; stress	pascal	Pa	$1\ Pa = 1\ N\ m^{-2}$
energy; work; quantity of heat	joule	J	$1\ J = 1\ Nm$
power	watt	W	$1\ W = 1\ J\ s^{-1}$
electric charge; quantity of electricity	coulomb	C	$1\ C = 1\ A\ s$
electric potential; potential difference; electromotive force; tension	volt	V	$1\ V = 1\ J\ C^{-1}$
electric capacitance	farad	F	$1\ F = 1\ C\ V^{-1}$
electric resistance	ohm	Ω	$1\ \Omega = 1\ V\ A^{-1}$
electric conductance	siemens	S	$1\ S = 1\ \Omega^{-1}$
magnetic flux; flux of magnetic induction	weber	Wb	$1\ Wb = 1\ V\ s$
magnetic flux density; magnetic induction	tesla	T	$1\ T = 1\ Wb\ m^{-2}$
inductance	henry	H	$1\ H = 1\ Wb\ A^{-1}$
Celsius temperature	degree Celsius	°C	$1°C = 1\ K$

Derived SI units with special names and symbols *(continued)*

Physical quantity	SI unit		Expression in terms of base or derived SI units
	Name	Symbol	
luminous flux	lumen	lm	1 lm = 1 cd sr
illuminance	lux	lx	1 lx = 1 lm m^{-2}
activity (of a radionuclide)	becquerel	Bq	1 Bq = 1 s^{-1}
absorbed dose; specific energy	gray	Gy	1 Gy = 1 J kg^{-1}
dose equivalent	sievert	Sv	1 Sv = 1 J kg^{-1}
plane angle	radian	rad	1[a]
solid angle	steradian	sr	1[a]

[a] rad and sr may be included or omitted in expressions for the derived units.

Physical Constants

Recommended values of selected physical constants[a]

Constant	Symbol	Value
acceleration of free fall (acceleration due to gravity)	g_n	9.806 65 m s^{-2} [b]
atomic mass constant (unified atomic mass unit)	m_u	$1.660\,540\,2(10) \times 10^{-27}$ kg
Avogadro constant	L, N_A	$6.022\,136\,7(36) \times 10^{23}$ mol^{-1}
Boltzmann constant	k_B	$1.380\,658(12) \times 10^{-23}$ J K^{-1}
electron specific charge (charge-to-mass ratio)	$-e/m_e$	$-1.758\,819 \times 10^{11}$ C kg^{-1}
electron charge (elementary charge)	e	$1.602\,177\,33(49) \times 10^{-19}$C
Faraday constant	F	$9.648\,530\,9(29) \times 10^{4}$ C mol^{-1}
ice-point temperature	T_{ice}	273.15 K[b]
molar gas constant	R	8.314 510(70) J K^{-1} mol^{-1}
molar volume of ideal gas (at 273.15 K and 101 325 Pa)	V_m	$22.414\,10(19) \times 10^{-3}$ m^3 mol^{-1}
Planck constant	h	$6.626\,075\,5(40) \times 10^{-34}$ J s

(continued overleaf)

Recommended values of selected physical constants[a] *(continued)*

Constant	Symbol	Value
standard atmosphere	atm	101 325 Pa[b]
speed of light in vacuum	c	$2.997\,924\,58 \times 10^{8}$ m s^{-1} [b]

[a]Data are presented in their full precision, although often no more than the first four or five significant digits are used; figures in parentheses represent the standard deviation uncertainty in the least significant digits.

[b]Exactly defined values.

The Periodic Table

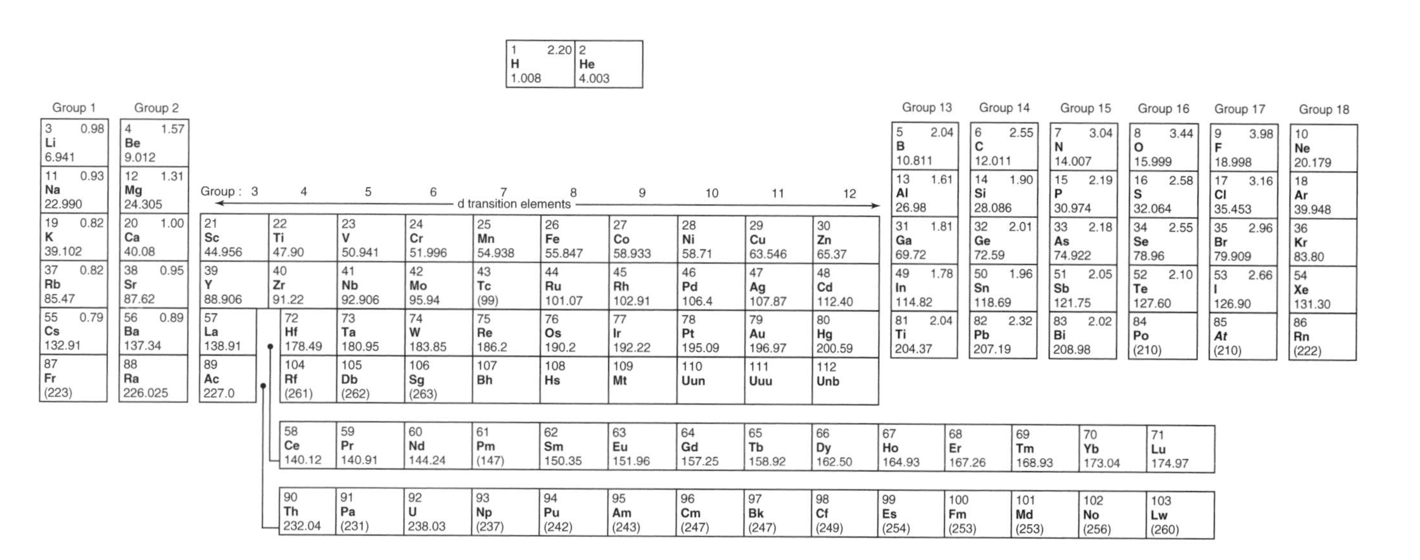

Key	
0.98	Pauling electronegativity
3	Atomic number
Li	Element
6.941	Atomic weight (^{12}C)

1 2.20 **H** 1.008	2 **He** 4.003

Group : 3 – 12: d transition elements

Group 1	Group 2	3	4	5	6	7	8	9	10	11	12	Group 13	Group 14	Group 15	Group 16	Group 17	Group 18
3 0.98 **Li** 6.941	4 1.57 **Be** 9.012											5 2.04 **B** 10.811	6 2.55 **C** 12.011	7 3.04 **N** 14.007	8 3.44 **O** 15.999	9 3.98 **F** 18.998	10 **Ne** 20.179
11 0.93 **Na** 22.990	12 1.31 **Mg** 24.305											13 1.61 **Al** 26.98	14 1.90 **Si** 28.086	15 2.19 **P** 30.974	16 2.58 **S** 32.064	17 3.16 **Cl** 35.453	18 **Ar** 39.948
19 0.82 **K** 39.102	20 1.00 **Ca** 40.08	21 **Sc** 44.956	22 **Ti** 47.90	23 **V** 50.941	24 **Cr** 51.996	25 **Mn** 54.938	26 **Fe** 55.847	27 **Co** 58.933	28 **Ni** 58.71	29 **Cu** 63.546	30 **Zn** 65.37	31 1.81 **Ga** 69.72	32 2.01 **Ge** 72.59	33 2.18 **As** 74.922	34 2.55 **Se** 78.96	35 2.96 **Br** 79.909	36 **Kr** 83.80
37 0.82 **Rb** 85.47	38 0.95 **Sr** 87.62	39 **Y** 88.906	40 **Zr** 91.22	41 **Nb** 92.906	42 **Mo** 95.94	43 **Tc** (99)	44 **Ru** 101.07	45 **Rh** 102.91	46 **Pd** 106.4	47 **Ag** 107.87	48 **Cd** 112.40	49 1.78 **In** 114.82	50 1.96 **Sn** 118.69	51 2.05 **Sb** 121.75	52 2.10 **Te** 127.60	53 2.66 **I** 126.90	54 **Xe** 131.30
55 0.79 **Cs** 132.91	56 0.89 **Ba** 137.34	57 **La** 138.91	72 **Hf** 178.49	73 **Ta** 180.95	74 **W** 183.85	75 **Re** 186.2	76 **Os** 190.2	77 **Ir** 192.22	78 **Pt** 195.09	79 **Au** 196.97	80 **Hg** 200.59	81 2.04 **Ti** 204.37	82 2.32 **Pb** 207.19	83 2.02 **Bi** 208.98	84 **Po** (210)	85 ***At*** (210)	86 **Rn** (222)
87 **Fr** (223)	88 **Ra** 226.025	89 **Ac** 227.0	104 **Rf** (261)	105 **Db** (262)	106 **Sg** (263)	107 **Bh**	108 **Hs**	109 **Mt**	110 **Uun**	111 **Uuu**	112 **Unb**						

58 **Ce** 140.12	59 **Pr** 140.91	60 **Nd** 144.24	61 **Pm** (147)	62 **Sm** 150.35	63 **Eu** 151.96	64 **Gd** 157.25	65 **Tb** 158.92	66 **Dy** 162.50	67 **Ho** 164.93	68 **Er** 167.26	69 **Tm** 168.93	70 **Yb** 173.04	71 **Lu** 174.97
90 **Th** 232.04	91 **Pa** (231)	92 **U** 238.03	93 **Np** (237)	94 **Pu** (242)	95 **Am** (243)	96 **Cm** (247)	97 **Bk** (247)	98 **Cf** (249)	99 **Es** (254)	100 **Fm** (253)	101 **Md** (253)	102 **No** (256)	103 **Lw** (260)

Index

General

Chemical and Biochemical Species